WATERS OF THE WORLD

SARAH DRY

WATERS
OF THE
WORLD

*The story of the scientists
who unravelled the mysteries of our
seas, glaciers, and atmosphere —
and made the planet whole*

SCRIBE
Melbourne • London

Scribe Publications
2 John St, Clerkenwell, London, WC1N 2ES, United Kingdom
18–20 Edward St, Brunswick, Victoria 3056, Australia

Published by Scribe 2019

Published in conjunction with The University of Chicago Press, Chicago

Printed and bound in the UK by CPI Group (UK) Ltd, Croydon CR0 4YY

Scribe Publications is committed to the sustainable use of natural resources
and the use of paper products made responsibly from those resources.

9781911617334 (UK edition)
9781925713145 (AU edition)
9781925693829 (e-book)

Catalogue records for this book are available from the
National Library of Australia and the British Library.

scribepublications.co.uk
scribepublications.com.au

in loving memory of Shirley Dry (1918–2014)
and
for Rob and Jacob

CONTENTS

1

INTRODUCTION

History can be cruel. Today, John Tyndall's grave in a quiet Surrey cemetery lies unremarked and his books largely unread. During his lifetime, he was a famous and famously controversial scientist who argued that nothing more and nothing less than molecules in motion could explain the deepest mysteries, from human consciousness to the origins of the universe. A gifted communicator, his lectures were standing room only. His books, merging physics and adventure, sold abundantly. He dined with the good and the great, among them Thomas Carlyle and Lord Tennyson.

Despite all this fame, the man whose passionate intensity fanned the fires of Victorian science is today almost forgotten. While the flame of his memory has flickered low, it has not been extinguished. In fact, in the past ten years Tyndall has begun to emerge from more than a century of near-obscurity. Thanks to work he completed in his laboratory in the late 1850s and early 1860s, on what he called the absorption of heat by water vapor and what we today call the greenhouse effect, Tyndall has gained newfound recognition as a so-called "father" of climate science. A handful of articles have appeared describing his discovery. A climate change research center at the University of East Anglia has been named after him, a major academic project is underway to edit his prodigious correspondence, and the first new biography of him to appear in more than sixty-five years has just been published.[1]

Tyndall has only recently resurfaced because the science of which he is being hailed as a progenitor is (somewhat paradoxically) itself so new. Little more than sixty years ago, climate was usually thought of as something that remained stable over time. Climatology was primarily a geographical science. Different places were understood to have different climates, and the job of the climatologist was to study not how those climates changed but what rendered certain regions distinctive. Their tools were descriptive and taxonomical rather than physical or mathematical. Climate science as a science of change rather than continuity (and distinguished from its older form, climatology, by the change of name) only emerged in the postwar period. When it did, it was the product of a blending of several distinct scientific disciplines. The journal *Climatic Change* was founded in 1977 with an editorial that made it clear that this was a science that existed almost defiantly between disciplines. Meteorology, anthropology, medicine, agricultural science, economics, and ecology were all encouraged to participate, though in fact the new interdisciplinary science centered around the physical sciences of the earth: oceanography, atmospheric physics, and glaciology, in addition to meteorology, with the important addition of the nascent field of computer science.[2] Before this interdisciplinary synthesis, the notion of climate change was an oxymoron.

The modern field of climate science, then, provides us with a challenge. How to tell the history of a new and self-consciously interdisciplinary discipline? Tyndall's increasing visibility as a "father" of global warming—alongside that of other progenitors such as Svante Arrhenius, Guy Callendar, and Charles Keeling—reveals a growing self-awareness on the part of climate scientists that history can be a tool to render this discipline more coherent. In these prehistories of climate science, success rather than failure is emphasized and crucial discovery "milestones" occur with reassuring regularity, like signposts on a journey to a known destination. Ideas tend to beget ideas, free of the complications of politics, economics, or nationalism. Histories told by scientists tend to be rose-tinted, but given the interdisciplinary origins of climate science, there is a perhaps even greater temptation than usual to pick and choose the moments, and in particular the discoveries, which make the most sense of the past, which generate a pleasingly direct line from the past to the present. Tyndall's modest re-emergence is part of a larger attempt by climate scientists to tell a singular history of a heterogeneous science.

The desire to draw straight lines through history is understandable, but these lines are almost inevitably misleading. John Tyndall was not the father of global warming in any meaningful sense. Though he helped confirm the special ability of water vapor and carbon dioxide to absorb heat as it radiated from the earth's surface, he never imagined that human beings might alter the climate on a planetary or even a regional scale. He was unconcerned about the carbon dioxide released into the atmosphere by burning coal. Nor did his research prompt his contemporaries to make their own research into human effects on climate. Nor, indeed, was Tyndall strictly speaking the first person to publish on the topic. An American woman named Eunice Foote beat him to it by three years. A "milestone" approach to telling the history of climate science misrepresents the complexity of its deeper history. Sometimes the result is to overemphasize the influence of a particular figure. More often, people and ideas that do not seem congruent with current scientific thinking drop out of this kind of history. The result is an impoverished understanding of the past as well as the present.

Tyndall did indeed help lay the groundwork for our contemporary understanding of the planet, and he more than merits a revival in the popular and scientific imagination, but he achieved his influence in more complex and contentious ways than the story of a singular "discovery" of greenhouse warming captures. What Tyndall achieved was to help change what it meant to study the earth. He did so in a passionately physical way—putting his body in danger and relying on his manic tendencies to enable him to focus on a problem to the exclusion of all else.

He used his physical energy to pioneer new ways to witness and to understand (the two always went together) the wonder of nature: its continuity. No substance better exemplified this continuity for Tyndall than water, a material he studied in all its manifestations with commitment verging on evangelical passion. "Every occurrence in Nature is preceded by other occurrences which are its causes, and succeeded by others which are its effects," he began his bestselling book on *The Forms of Water*. "The human mind is not satisfied with observing and studying any natural occurrence alone, but takes pleasure in connecting every natural fact with what has gone before it, and with what is to come after it." Tyndall invited his reader to join him in tracing a river to its source, to follow it beyond its many tributaries and up into the atmosphere itself, from which it had fallen as rain. To produce that rain, Tyndall continued, water vapor must have been evaporated, via the action of heat, from the ocean into the atmosphere. This landed him at the ultimate source of all movement on earth. "Is there any fire in nature which produces the clouds of our atmosphere?" Tyndall asked rhetorically, before answering triumphantly that "by tracing backward, without any break in the chain of occurrences, our river from its end to its real beginnings, we come at length to the sun."[3]

And here, with the sun's heat, we arrive at the deeper value in studying water. Constantly transmuted by the energy of the sun, water provides the mechanism by which energy flows through the landscape. In his insistence on continuity, made gloriously manifest in the substance of water, John Tyndall offered his own, Victorian version of interdisciplinarity—a way of thinking across scales of time and place as well as linking what were, even in the nineteenth century, the increasingly divided spheres of the arts and sciences. As such, he provides a window onto what it meant to study the earth and what we have come to call its climate long before a "science" of such matters existed, much less anyone had imagined we might perturb the global climate. Tyndall is a gateway into another way of understanding the history of climate science. His passionate engagement with water as a medium for studying what he called the continuity of nature inspired this book. But rather than revealing the wonder of nature alone, I hope to introduce a wonder of the human sort.

In this book, water traces not the flow of energy but the flow of human activity and thought, from the work of Tyndall and his contemporaries to those scientists who helped shape the earth sciences in the twentieth century. This brings the science alive and it also helps solve the conundrum of how to tell a history of climate science that is faithful to its multidisciplinary nature. Climate science clearly has implications that extend far beyond the boundaries of science itself. So, too, the history of how we've come to understand the planet should matter not just to climate scientists but to all of us. In the life and work of scientists from the past lies the opportunity to understand the origins of our own way of seeing the world.

We are currently engaged in a global effort to understand simultaneously how our planet works and how we have affected and continue to affect it. Some of the tools we use to do so currently go by the name of climate science. This book proposes to tell the stories of a few key individuals in the history of the sciences of water in order to illuminate the broader history of human understandings of the planet over the past 150 years. In doing so, I hope to reveal not only continuities but discontinuities between the present and the past. We are inheritors of both more and less than we know.

It is now quite easy to see how human beings have made their mark on even the most remote places on Earth. Floating islands of plastic blight the remote ocean. Trash litters distant Alaskan coasts. The invisible rising presence of carbon dioxide in the atmosphere is everywhere. Ice sheets are, as the Danish ice-core scientist Willi Dansgaard memorably called them, the ultimate frozen annals, recording enormous spans of time in their compressed layers of ice, trapping past atmospheres, dust storms, and volcanic eruptions. These icy archives are just one of the records from the past that we have learned how to read. Lake and ocean sediment, underground stalactites, and tree rings also preserve the history of the earth, and of human presence on it. They tell stories of increasing human impact, as well as older histories that predate us.

These records are important and have much to tell us about the past, and, because they help us generate the climate records against which our models can be checked, they enable us to try to peer into the future as best we can. As important as these material records are, there are other records of the past that are equally important in helping us understand our climate today but which remain largely inaccessible and underutilized despite the urgent need to understand climate as comprehensively as possible. These are not the physical traces of past climate but the imaginative traces of past understandings of climate. Our imaginations have shaped our understanding of the planet, and our scientific imaginations have shaped our understanding in particularly crucial ways. In his book *Landscape and Memory*, Simon Schama writes that "even the landscapes that we suppose to be most free of our culture may turn out, on closer inspection, to be its product."[4] We can tell this is true because our attitudes toward landscape change over time. Mountains, once considered horrific, have come to be seen as paradigmatically beautiful. The first settlers in the Americas experienced the landscape as an empty and desolate wilderness, both spiritually and materially vacuous. Today we might call such a landscape magnificent and full of life. Each of us responds to the landscape around us according to the cultural habits we've acquired without noticing. We each have our own taste, a preference for coastline or valley, cityscape or farmland, but these individual differences play out across a backdrop of shared attitudes that change only gradually over time (though they may vary quite dramatically across different cultures).

I would add to Schama's elegant formulation that it is especially those landscapes that seem to be most free of our culture that show its influence. The imaginative understanding of wild spaces that in the West are perceived as lacking human interference have much to teach us. Prime among these are places such as the upper reaches of the atmosphere, the depths of the ocean, and the icy heart of a two-mile-thick ice sheet. Scientific culture works like any other culture to break down and materially change the substrate upon which it lands. This can be seen as a positive outcome: the discovery of order in a place of apparent chaos. It is also possible to see not clarity but distortion at the interface between science and the natural world. We see what we want to see, in other words. More strongly still, we might say that in the act of observing, we change the thing being observed. A glacier is affected only slightly by the presence of people upon it, but their study of glacier motion, their passion for knowing the way the glacier moves in terms of inches per unit of time, obscures and elides countless other ways of knowing the glacier—as an object of beauty, of terror, of passage (as it was to the locals), of uselessness, of unpredictable destruction (in the form of crevasses, avalanches, and dams bursting). By suggesting that truth can be found in the science of glacier motion, for example, Tyndall and his contemporaries also contributed to the idea that some truths about glaciers—for example, how quickly they move—are truer than others. It is a poignant irony that Tyndall himself was such a passionate advocate for the emotional truths to be found atop glaciers, since he played such a central role in transforming them into sources of "merely" physical truths.

There are many ways of going about the task of uncovering the imaginative assumptions that we make about remote places. Historians of these cultural attitudes rely upon a variety of texts to draw them out. Literature is a good source for finding reflections, echoes, and elaborations of such cultural themes. So are painting, photography, and drama, arts that both reflect and enhance the assumptions of the culture from which they emerge. Scientific writings were also once good sources for charting the shifts in how people have felt about the natural world. Until the late nineteenth century, most scientists wrote books for everyone to read; by rereading those books now, we can see what kind of public knowledge there was at the time and, by working a bit harder in the archives, we can try to get a sense of who read these books and what they thought of them.

More recently, most scientists stopped writing for a general public. Instead of writing articles in popular magazines, they began to publish technical articles for their scientific peers in expensive and hard-to-access scientific journals. Where scientists might once have shared a long narrative about the process of gaining new insights— say the expedition to study the motion of glaciers, or a voyage to South America—those stories have been largely removed from modern scientific papers, reduced to the terse terms of the Methods section. That is, at least, what happens in public. In private, stories of suffering and triumph in the field circulate still, at conferences, via email, and over cups of tea and pints of beer. The desire to share experiences, to brag, and to caution is not going to fade away anytime soon. The difference is that it is harder for the public to eavesdrop on them.

In an attempt to redress this loss of access into the experience of science, what follows are stories of scientists doing science. They are not stories of made-up things, or purely imagined or projected understandings of the earth. They are the stories that reveal hard-earned skills in observation, measurement, calculation, and description, and the careful construction and skillful deployment of instruments that are made useful through a combination of discipline, training, and social convention.

The transformation of a scattered array of fact, theory, observation, and experiment into something that can be called global knowledge is an example of the necessary sleight of hand that animates all science whereby certain bits of understanding—a key experiment, a set of measurements that underlie a mathematical abstraction—are taken to justify insights into the behavior of nature everywhere. But it is especially important in the case of the planet, which is, after all, a singular unit, the only one of its kind, which we have come to think of as self-evidently whole. It is one job of this book to show that the self-evidence of that wholeness is a very hard-won result, the outcome of the work of many scientists working at many different locations at many different times.

If much gets put into global knowledge, it is equally important to remember how much gets left out. Just because we are all on this planet together does not mean that each voice can speak equally loudly. That is obvious in the realm of politics, but it is less obvious when it comes to telling the histories of science, where the story of creating global knowledge is often taken for granted as a story of unmitigated progress. This narrative of progress is, to a greater or lesser extent, the melody by which all histories of science tend to get sung. Science is understood, fundamentally, as a progressive human endeavor. And in some senses it is. But in other senses, it is equally a process of elision, excision, and exclusion.

Big ideas are often invisible, so influential that we can no longer imagine looking at the world in any other way. We think we are simply seeing things the way they are. The idea of the earth as a global system of interconnected parts is a case in point. It is so basic that even those who still question the reality of anthropogenic climate change share it. The idea that there is a global climate is rarely a topic of debate (though when you start to think about this, it is hard to say what—or, more to the point, where—such a climate might be). The debate has hinged instead on whether the global temperature is rising or falling, or, as the rise becomes increasingly hard to deny, what the future will look like. The idea that there is a climate system, a set of natural features that are interrelated and function at the global scale—well, that has come to seem obvious.

What is the history of this "obvious" fact? Many point to the famous "Blue Marble" images taken aboard the 1972 *Apollo* 17 NASA mission, images that gave us our first glimpse of the planet as a whole. Seeing Earth like this, the story goes, was a revelation. We grasped, finally and instantly, the fragility, the uniqueness, and the interconnectedness of everything on the planet. The vision of the dazzling Earth rising above the barren surface of the moon did give a big boost to the burgeoning environmental movement. But we were already primed to see Earth that way. The space race was more a product of the previous successes of a global vision of the planet than it was a producer of it. Long before *Sputnik* and the *Apollo* missions, scientists had helped craft a vision of Earth as a globally connected object out of countless investigations into the physical complexity of our planet.[5]

Gaining a better grip on the breadth of knowledge that constitutes climate science today is essential for understanding what we know and what we don't know. The tendency to judge climate science by its predictive abilities has serious consequences for how we make decisions, as citizens and nations, in the face of uncertainty about the future. Our contemporary expectation that climate science can and should make predictions about the future indicates the long shadow of that old "pattern" science, astronomy, still extending over us today. But within what is too often monolithically referred to as climate science, there are many different methods for generating knowledge. These methods are sometimes called subdisciplines. For the story of global climate knowledge, they include geology, climatology, meteorology, atmospheric physics, glaciology, and computer science. In order to understand how our knowledge of the planet has come to feel global, we need to understand how these disciplines within the larger body of science have come to seem interrelated. The history of our knowledge of the planet is necessarily the history of the disciplines (and all their associated practices, pedagogies, instruments, techniques, and social structures) that have generated that knowledge. To create a singular global climate, in other words, it was necessary to forge a unified climate *discipline* out of many parts in just the same way that it was necessary to find ways to bring what had previously been disparate pieces of knowledge—of this place, say, or this type of object—together.

To understand the nature of climate science (understood broadly) requires going back to the particulars out of which it has been generated. This means places and people. My previous two books are biographies (one about Marie Curie, the other about Isaac Newton's manuscripts), and my instincts are biographical. So that is what I have chosen to do here. People, not water, are the true subjects of this book. These people are scientists. The oldest of them was born in 1819. The youngest was born in 1923. I watch the planet with their eyes, take a journey through the past with them as companions and investigators, as explainers and exclaimers. This investigation of watery things is, then, very much a grounded one, planted firmly in the personal experiences of a remarkable group of thinkers.

I begin in the 1850s with the first attempts to measure changes in climate and weather simultaneously and at a global scale—the beginning of both modern weather forecasting and climate science. It is here that I also trace pioneering studies into the importance of the atmosphere in regulating the climate—at a time when no one dreamed that human beings might affect the temperature of the earth as a whole. Yet this was also the time when the new science of thermodynamics seemed set to crack open untold mysteries not only of the earth but of the entire universe. New equations could explain the behavior of molecules statistically. It remained to be seen whether these equations could also explain the movement of molecules in the real world of glaciers, clouds, and water vapor.

In the 1850s, it was glaciers that threw up the biggest challenge of all to scientists hoping to explain their motions, and, by extension, the past and future of the earth's climate. Although the ice ages are now a taken-for-granted fact, they once seemed both real and inexplicable, a puzzle of mind-boggling extremes to be solved on a global scale. John Tyndall sought answers to these deep questions of time, movement, and decay surrounded by the deadly, searing beauty of Alpine mountain glaciers and, on his return to London, in the confines of his basement laboratory. His findings on how heat acts on ice and water vapor reveal an obsession with energy, with dissipation, and with the past and future of the planet.

In 1856, Charles Piazzi Smyth, a Scottish astronomer and scientific traveler, tried initially to subtract, or to erase, the presence of water vapor from his astronomical researches atop a high volcanic peak in Tenerife, one of the Canary Islands. Later, he hoped that the study of water vapor with a powerful and highly portable new instrument could help make weather prediction safer, more respectable, and possibly even successful. He failed in that endeavor, and tarnished his reputation with a passionate defense of the idea that the British measurement system had been divinely inscribed in the Egyptian pyramids. Finding himself outside the scientific establishment, he sought solace in a peculiar blending of religious and scientific witness, a photographic cloud atlas he attempted to assemble alone, in his final years as an isolated, embittered, but always reverential scientific pilgrim.

Both Tyndall and Piazzi Smyth strained to contribute to a predictive science that could account precisely for the actions of water—of the movement of glaciers, of the action of water vapor, of the formation of clouds and the falling of rain—and both welcomed the feeling of mystery and wonder that accompanied their investigations even as they attempted to describe the world dispassionately. These men experienced the inherent contradiction of these two positions with a passionate, even visceral intensity. Their stories capture the torment this Victorian generation experienced as they attempted to reconcile science's potential to reveal the hidden structures beneath the wild confusions of the earth's environment with the loss that might accompany such revelation. Would the gain in understanding compensate for the forfeit of mystery? In many ways, Tyndall and Piazzi Smyth were members of the last generation of scientists for whom such an existential struggle had an acceptable public face. They published books that invited general readers to feel their fear, wonder, and awe as they encountered sublime phenomena such as cloud forms and majestic glaciers. And then they tried to reduce these phenomena to numbers, equations, and theories that could not merely explain but also predict the most intimate details of what had previously been, almost by definition, ineffable.

The story of Gilbert Walker, a preternaturally talented English mathematician, provides a transition between the nineteenth century, when individuals could still express scientific ideas in books for the general public, and the twentieth century, when dry scientific papers replaced the dramatic travel narratives written by men like Piazzi Smyth and Tyndall. When Walker became director of meteorological observatories in India, many believed that the key to unlocking the secrets of the monsoon rainfall, upon which millions depended (and still depend) to sustain their crops, lay in the cycles of spots on the sun. Walker's statistical inquiries, made possible by the access to weather data gathered via imperial networks and by the hard work of local calculators employed by the British government, destroyed the cherished hopes of the sunspot theorists. In place of the congruent, or coherent, harmonies of sun and Earth, Walker offered a statistical discovery of amazing scope. His calculations indicated a connection (actually a tele-connection) between the monsoons in India and pressure and temperature halfway around the world. Walker named the phenomena of linked meteorological phenomena "world weather," and, more specifically, the one affecting India he called the Southern Oscillation. Unlike

Tyndall, who was committed to demonstrating the links between physical phenomena, Walker's scientific insights were purely statistical. He could not explain how pressure in the west Pacific affected rainfall in the Indian Ocean; he could only say that it did. (In fact, it was another forty years before the physical links that drove the Southern Oscillation could be explained.)

A golden age of physical oceanography and meteorology was initiated by a bolus of funding and urgent practical need for information about air and water during World War II. It continued in the Cold War for decades thereafter. This was a time of big pictures built on remarkably simple models, characterized both by new kinds of international cooperation and by the tensions of realpolitik. Henry Stommel was a young man in 1948 when he published a paper explaining why every ocean basin in the world has a fast-running current on its western side. His fruitful thinking led the way for a new generation of oceanographers who showed that the ocean was moving in a much more complex and energetic way—on a multiplicity of time and spatial scales—than previous generations had imagined. In so doing, Stommel set the stage for an ocean characterized largely by its turbulence rather than its stability and a new way of doing experiments in the ocean that required large-scale and long-term cooperation, something Stommel himself intensely disliked. At roughly the same time, Joanne Simpson investigated how the relatively small-scale dynamics of clouds could drive atmospheric—and oceanic—circulation on planetary scales. She also sought new ways to do science—using instrumented aircraft and canny cooperation with government agencies to experiment on clouds, and even hurricanes, by seeding them. This work on weather and climate modification took place against a backdrop of anxiety about the threat of attack from the Soviet Union. These water stories show how a connected Earth can be a vision of war just as much as of peace.

Individual scientists were both pawns and hustlers in this worldwide game of scientific chess. When the Danish physicist and meteorologist Willi Dansgaard realized that the new mass spectrometer he had access to could be used to sort water molecules by weight, he was following his own private intuition. But to follow his insight to its fullest conclusion, he had to convince the largest and most powerful national and international scientific (and sometimes military) agencies to give him access to technologies he would never otherwise be able to afford. The story of ice cores and the history of past temperatures (or "palaeothermometry") is a story of individual cleverness, tenacity, and diplomacy played out against the backdrop of the Cold War. Dansgaard's contribution helped change our understanding of past climate and laid the foundation for the first glimmerings of what would become our contemporary awareness of global climate change. But, as I show, the assumption that one part of the planet—in this case, the northern Greenland ice sheet—could umproblematically speak on behalf of the whole turned out to be, in important particulars, inaccurate. The *idea* of global changes captured in ancient ice turned out to be more important than the facts recorded in those cores.

For all the triumphs these scientists enjoyed, their stories are also threaded with loss. The loss is often personal—one scientist cannot make human relationships work, another suffers a crippling nervous breakdown—but it is also existential. Global knowledge of the kind these scientists create has often been prompted by questions about changing conditions on the earth. New knowledge also prompts new questions about our relationship with what we know, how we know it, and how we should feel about that knowledge. The role of mystery, ignorance, and wonder in the pursuit of science and the implementation of its findings remains as important as ever, though we have lost the Victorians' readiness to acknowledge this fact. In its attention to the role of sentiment, awe, and longing, this book is as much a history of emotion as it is a history of science.

While we are today preoccupied with our own anxiety about the effects of human-induced climate change, previous generations have voiced different concerns. In the light of our modern awareness of global warming, these early investigators constantly seem to be getting it wrong, to worry about irrelevant things, to miss what seems to be right in front of them. But of course it is we who are getting it wrong if we look backwards only to find what we think we know today. Instead, I revisit the passionate commitments of these scientists from the past to better understand what it was that motivated them. These include Tyndall's melancholic conclusions on the implications of the second law of thermodynamics, the millennial anxiety caused by nuclear weapons testing in the 1950s and 1960s which released radioactive elements into all parts of the water cycle, and the flip-flopping worries about the effects of widespread cooling and global warming in the 1970s. Less dramatic but perhaps even more fundamental is the shift from relatively simplistic models of ocean and atmospheric circulation to models that are fundamentally chaotic. Such a loss of even the possibility of certainty casts the crisis around climate change further into the shadows of political contestation. If science cannot give us certainty, goes the argument, what can? Will we need to invent new kinds of knowledge, new ways to know what Earth is? Or will we need to let go of the idea of certainty and embrace a changing planet?

In Tyndall's book on *The Forms of Water in Clouds and Rivers, Ice and Glaciers*, he talks his reader (to him readers were listeners, and texts like stories frozen on the page) through a world he'd explored for many years and come to love. It is a water world full of such energy and movement that it feels alive, though living creatures are notably absent. Banners of cloud stretch, crevasse-slashed glaciers buckle, and crystalline lake-ice cracks. Above it all swoops Tyndall himself, taking in hand an imaginary listener who, he admits, has become to feel so lifelike to him that by the end of the exercise he feels real affection for the "abstract" boy. Tyndall is like that: He can't help bringing things to life, be they imaginary boys or frozen landscapes.

The book you hold in your hands puts water to a similar purpose. It uses stories of water, as Tyndall did, to animate the most inhospitable and lifeless spaces on Earth—the deepest recesses of the open ocean, the vast ice sheets of Greenland and Antarctica, the water vapor that suffuses the farthest reaches of the atmosphere. While Tyndall made his imaginative voyages in order to share with readers the wonder of the natural world, the wonder I want to share is historical. How have our water stories changed over time? What do these changes tell us about what we know (and what we think we know) about the planet today? How can they help us prepare for a future that remains, as it must, uncertain?

2

HOT ICE

We can begin anywhere. As John Tyndall has made abundantly clear, all things are connected, and to pull a thread from one part of his life should reveal the warp and weft of the whole. So let's start here, in the place he most loved to be: on the side of a mountain in the Alps. It is December 1859 and John Tyndall, aged thirty-nine, is walking up a mountain. He puts one foot in front of the other in an easy rhythm. A wiry man, he disdains eating while he is on the move. His friends call him the goat. The sky is radiant blue, deeper and clearer than it ever is in London. The peaks and pinnacles are craggy ruins. The glaciers are pristine, the snows of winter blanketing the pocks and streaks caused by summertime melt. The clouds are almost but not quite more than Tyndall can describe.

He carries very little with him. A notebook, a flask of tea, a hard biscuit jammed in his back pocket. A walking stick with his name engraved in a charred and wiggly line by his friend Joseph Hooker with a pocket lens and the sun's heat. A pair of good boots and a kerchief tied around his neck against the cold. Reluctantly, he has employed two guides and four porters to accompany him.[1] Though they carry his heavy gear, he'd rather travel solo. The mountain is a balm to him in proportion to his solitude and the level of danger he faces. Danger is cheap on the mountainside, available to him at half a pace's shift to the left or right. A loose rock, a long slide into oblivion. Other dangers are available, too. The seductive embrace of the cold if taken unawares by a snowstorm. Collapse brought about by overexertion in the thin air. So far Tyndall has been lucky, and he has been careful. He does not seek thrills, only catalogues them as he climbs, his eyes darting this way and that, from trail, to precipice, to the skies overhead. His mind is busy with this cataloguing, and this frees him from thinking of other things. His life back in London as professor of natural philosophy at the Royal Institution. The unfortunate tendency he has to dispute. The lonely whirring of his mind during an insomniac night, followed the next day by a cottony head, dry eyes, and a racing heart.

He puts one foot in front of another and watches the skies. The morning clouds clear so evenly, it is as if a dial has been turned by a deliberate hand. Tyndall knows that the water has not disappeared, only become something else, an invisible vapor rather than a spray of fine droplets. He stops for a rest and kicks a rock, just to see where it goes. It falls close to the mountain and then hits the side to make a huge bouncing arc into the thin air, before falling back against the mountain and bringing down a rain of smaller rocks that Tyndall can hear, like gravel on a roof. There, he's done his bit to bring the mountain down one tiny increment farther. If he thinks too long on it, the decay of the mountain makes him melancholy. He cannot help unspooling his thoughts further into a cold and dismal future, to a time without humans, without life and without even the heat of the sun.

His somber meditations do not long dominate his thoughts. Just as often he finds occasion for a cheering, even valedictory, feeling of unity with the forms of water around him. Traveling to the glacier on Christmas Day, he'd happily embraced the unity of the nature that surrounded him. "The heavens mostly grey," he recorded in his journal, "a clammy vapour overspread the lake, and the red of the eastern sky had dusky cloud streamers drawn across it in different directions. . . . The hoofs clinked merrily upon the frozen road: right and left we were flanked by snow, but in the middle of the road, this was pressed to hard ice. As the valley narrowed, and the mountain walls came nearer this softened, and at some places had melted almost entirely away. As day advanced, the clouds vanished and a fine blue dome stretched overhead."[2]

Soon after arriving, Tyndall and his guides headed for a rough chalet called Montanvert, close to the Mer de Glace, the glacier where they planned to make measurements. Tyndall had been here before, with the same aim, but never in winter. Soon snow was falling, astoundingly fast and thick. The men made it to the building where they would shelter, and Tyndall lay listening to the wind howling through the night. In the morning, he watched as the red light of the rising sun hit the clouds that fringed the steep ridge rising above the glacier. He recalled Tennyson's "eternal" phrase, "God made himself an awful rose of dawn."[3] For a moment, the mountains burned like a pair of torches, and then the sun rose fully and the day began.

With daylight, the men started the work that had brought them here. Tyndall arranged his theodolite, a surveying tool for measuring locations precisely, in heavy snow, while the men advanced onto the glacier. Once in position, they set stakes in the snow along a line determined by Tyndall and his instrument. Heavy snow swirled across the glacier. Tyndall was able to communicate with the men only during brief moments when the air cleared. For the most part, all was whiteness and wind.

As the day continued, the quality of the snow changed and the flakes became blossom-like. They fell thickly on his coat, "soft as down." Such prodigal beauty seemed to Tyndall a rebuke to humanity's overweening pride. What could human beings matter to Nature if she is content to put on such shows where none can see them? It is typical of Tyndall that even as he considered his own insignificance, he recorded his unstoppable appreciation of the beauty.

After three hours in the snowy blizzard, Tyndall and his men managed to complete the measurement of a single line of stakes across the glacier. They returned the next day to record the distance they had traveled from their locations the day before. The weather remained awful, but Tyndall managed to make the measurements in the brief moments when it cleared. By midday, the work was complete. As he set back down the glacier, he turned and observed the line of stakes with a philosophical eye. "I knew of course that I had set them there, still the idea of intelligence and order that they suggested in the midst of that scene of desolation was pleasant"; he later recorded, "it seemed like the perception of law in apparent confusion."[4]

The air atop the glacier was fresh and sere, as if all the moisture had been evacuated from it. Before he turned to the absorbing business of the descent, which required every ounce of his attention to be kept focused on the ice beneath his feet, Tyndall took a last lingering look at the great snow-filled valley and wondered at the ages it had taken for the miniscule action of molecules upon molecules to carve out the landscape before him. How long it had taken for these vast ice fields to accumulate was more than he, or anyone, could say, though some had tried. In a good year, some forty centimeters of new snow was laid down. How much was compacted by the weight of the snow that came after it, leaving a narrow sliver to represent an entire year of steady accumulation of snow? And how many of those tiny bands made up the bones of the glacier? It was impossible to know. Worse still, the glacier produced its own excretions, the freshwater that emerged from its lowest depths like a spring from the ground. This water was like a liquid breath, evidence that the glacier lived. It also destroyed the chance of a calendar written in the ice, since it was always the very oldest ice that had been slowly melting. And who could say for how long?

In the search for certain fundamental laws of nature in the chaos of the Alpine glaciers, Tyndall felt more alive than he ever did in London. The danger of the ice focused his mind, forcing him to think only about his next step and the one after that. Somewhat paradoxically, considering how many had already visited before he had, being on the glacier also allowed him to see himself as being alone in his scientific enterprise. If he was singular, he had a better chance at winning the prize, of explaining how sheets of ice could move across the earth. When he thought of the race to be first, the image of the virgin field of snow came to his mind, the glacier a pristine whiteness, whose "billows rose steep and pure and sharply crested."

FIG. 2.1. John Tyndall in 1857, around the time of his first glacier investigations.

Of course he was alone neither literally, since he had his porters and his guides, nor figuratively, since he carried with him all the weight of previous theories that had come before, and all the fellow scientists who had put them forward, men who were simultaneously his imaginary collaborators and fellow combatants in the fight for priority, for influence. He selected his approach, and marshaled his resources accordingly. What he chose to do on the mountain was to undertake an open-air experiment. He would not simply look and record what he saw, as the geologists did, but would use his stakes, his theodolites, and his obedient assistants to subject the glacier to a particular kind of investigation—an experiment— more suited to physics than geology. As experiments go, this one was relatively crude. Measurements were made in yards rather than millimeters, and time measured in days rather than milliseconds. But it was still an experiment. He was trying to answer a specific question: How fast does the glacier travel at different locations? He would use the measurements he had stolen from the blizzard to support what he called his "theory of glacial movement" (others have different ideas). From a comparison of the locations of the stakes on the two days, he determined that the glacier had shifted 15.75 inches in that location. The same portion of the glacier, he knew from previous observations, had moved a little more than twice as fast during the summer.

The speed of the glacier would give an indication of the mechanism of its motion. What Tyndall had come to the Mer de Glace to do specifically was gather evidence with which to formulate and bulwark a theory that could explain all of the surprising facts about glacial motion. In the process, he hoped to claim victory for his theory over that of the religiously and socially conservative Scottish geologist James David Forbes, eleven years his elder and the man he considered to be his main opponent.[5] Forbes had already said that ice moves like a viscous substance, but Tyndall wanted to show that this was merely an observation, not a theory. In itself, according to Tyndall, Forbes's set of observations explained nothing. Worse, Forbes's theory actively obscured the true nature of glacier motion. Tyndall wanted to show not merely what the ice *seemed* like (treacle or honey) but how it *actually* moved. Tyndall had chosen his destination for this winter expedition carefully. He came to the Mer de Glace because he wanted to be in the place upon which "the most important theoretic views of the constitution and motions of glaciers are based."[6] Many other scientists had stood upon this same glacier, the largest and most accessible in Europe, making their own observations and formulating their own theories. What all of these men had been wanting to know was the precise mechanism by which ice moves. Following in their footsteps was the only way to surge ahead. Tyndall must witness the same phenomena with new eyes, and submit them to new experiments, to prove that his understanding was superior to those of the men who had come before. If he went to another glacier and made his experiments, critics could

FIG. 2.2. James David Forbes, Tyndall's adversary in the matter of glacier motion, traveled to the Alps during the 1840s to make measurements on the ice.

always argue that the phenomena in question were different, that the results didn't apply. But if he did them in the same place and showed that his account of glacier motion was superior, then he would have vaulted to the front of the pack.

Until recently, it simply hadn't occurred to anyone, with a few important exceptions, to consider whether the ice could move, much less how it might accomplish such a thing. Those exceptions were the local people, mostly shepherds, who lived and worked in the mountains amid the glaciers. They noticed, year after year, the subtle and sometimes not-so-subtle changes in the glaciers that indicated movement. They saw the scratches on the sides of steep mountain valleys, the long piles of rocks discharged at the foot of the glacier.

FIG. 2.3. The Mer de Glace, in the French Alps, where John Tyndall and James David Forbes both made measurements on glacier motion.

They lived with the occasional catastrophe, too, when the ice dams that preserved glacier lakes burst, sending huge amounts of water and terrifying blocks of ice tumbling through otherwise serene mountain valleys. But these people were not natural philosophers. They kept their thoughts to themselves, and it occurred to none of the small cadre who called themselves "gentlemen of science" to ask them what they thought.

It was only in the 1830s that the Alps started to matter to those who did not live and work there. Only then, when the mountains started to become more than a curiosity to men of science in places like London, were the seeds sown that gave Tyndall's idea of journeying to the Alps an urgency that he was unable to resist. To a surprising degree, it was the hot and heavy world of industry, commerce, and trade that turned an obscure and icy corner of Europe into a major site of scientific inquiry. The tracks laid down by hungry railway companies as they sliced across Britain and the mines that dug ever deeper to find coal brought more and more secrets of the earth's crust to light. These newly exposed strata and fossils raised increasingly hard-to-ignore questions about the earth's history. The people who spent time scrambling over these new rocks with hammer and magnifying lens at the ready were concerned, first and foremost, with coming up with a good story to tell about the earth's past. But the railway and mining companies also stood to make vast amounts of money on the basis of the advice these men could give them about where the earth's hidden riches might lie. Soon, the rock-scramblers had a collective name for the study of the earth's history—geology. These geologists looked to the Bible, with its compellingly dramatic narrative of the universal flood, as a resource that could guide and check their studies, but most were happy to read scripture metaphorically, translating a biblical day or year into thousands or even millions of years, if need be. Most important was the way that the intensely human Bible story—studded with contingent events— could still serve as an exemplar for a chronology of the earth's history that was, in literal terms, vastly longer than anything it contained within it. The idea that the earth had a history that was separate from, and far longer than, human history was new and unfamiliar. But the structure of this astoundingly long history was not new at all. When Tyndall and his peers inherited this way of thinking, they were inheriting a way of looking at the earth that was rooted in the most familiar stories of all. This kind of earth history looked like the human history recorded in the Bible, a history marked by twists and turns, by the sense that, but for the intercession of chance events, things could have turned out differently.[7]

In addition to being provoked by the new evidence from below the earth's crust, geologists in the early nineteenth century also started to look differently at things that were lying on the earth's surface, which they had previously either ignored, discounted, or simply failed to see because they were not looking for them. Erratic boulders, of a type of stone alien to the landscape in which they were found, had always puzzled locals and provoked geologists to try to explain how they came to be where they were. Strange deposits were often found near these erratic boulders, a layer embedded

FIG. 2.4. Tyndall, center right with beard and hat, stands with other members of the Alpine Club outside their club room in Zermatt. From Edward Whymper, *Scrambles Amongst the Alps* (London: John Murray, 1871).

with stones of every shape and size, arranged in no particular order. These disordered, unlayered beds were mystifying to geologists whose main tool for analyzing the structure of the earth was the comparison of fossils that had been laid down in stable strata.

The fragmentary nature of these so-called drift deposits, as well as erratic boulders that seemed to be scattered across the landscape without apparent order, posed a substantial challenge to the geologists trying to explain what caused them. For a long time, the leading explanation was a great flood, or series of floods, powerful enough to displace and transport even very large stones hundreds of miles. But the idea of a global deluge was too violent and too improbable for Charles Lyell, the pre-eminent geological thinker of the day. By 1835, Lyell had developed an explanation more compatible with the idea that geological change happened gradually as a result of causes that could be seen acting today. The unusual drifts, he said, were best explained by the existence of a huge sea, created by a gradual but dramatic lowering of the continents, which had once covered most of the globe, across which floated myriad icebergs, laden with a cargo of stones and clay. As the icebergs melted, Lyell imagined, they deposited their rocky loads on the seafloor. Because the movements of the icebergs were jumbled, their deposits were similarly chaotic. This theory had the merit of justifying the failure of geologists to bring clarity to their study of these drift materials. Clarity was elusive, according to the iceberg theory, because the mechanism of deposit was the jolly confusion of floating icebergs. Since icebergs had been seen in recent times, during a comparatively warm climate, his theory had the merit of not requiring a massively different climate in the past. "Adoption of this theory of ice-drift," reassured Lyell, "does not of necessity require us to assume the former existence of a colder climate than that now prevailing in North America."[8] It made Lyell distinctly uncomfortable to imagine a past climate that was dramatically different from that of the current day.

Growing up, Tyndall heard of frequent voyages in search of the fabled Northwest Passage that could funnel British ships through the narrow inlets and icy seas of northern Canada to the other side of the world. Expedition narratives told of bergs so massive that ships floated alongside them for weeks at a time. These stories strained credulity at first, but, once authenticated, they were the perfect grist for Lyell's theory. Indeed, Lyell drew liberally on reports of towering icebergs and massive ice sheets from expeditions to Canada and Greenland to construct a theory of sufficient grandeur to explain the widely dispersed geological puzzles. Sightings of icebergs as far south as forty degrees north suggested that very mild climatic change in the past could have produced the kinds of deposits that geologists were trying to explain. In 1819, William Parry brought back tales of enormous bergs that dwarfed the sailing ships—one estimated to be some 860 feet high (including the underwater portion). In 1822, whaling captain William Scoresby headed the first English expedition to lay eyes on the eastern coast of Greenland. The vast island backed by coastal mountains seemed to be topped by an ice sheet of unimaginable size. News came from the south as well, from ships in the Antarctic that sailed through a watery landscape strewn with icebergs. These reported visions—treated as facts thanks to the authority of the men who made them—fired the imagination of writers, poets, and playwrights. Stories of the ice helped bring the ice closer to home, to domesticate the ice for Tyndall and

FIG. 2.5. Louis Agassiz, the Swiss geologist who promoted the bold new theory of the ice ages in the 1840s.

his contemporaries. Ice became part of popular culture. In 1816, a young Mary Shelley framed her gripping tale of a scientist's creation of new life with a cautionary story of Arctic exploration and loss. Ice, in the early part of the century, was a source of new sensations that titillated and terrified the public in equal measure.

In 1840, Louis Agassiz made a bold suggestion that changed forever the way that everyone—scientists and the public—thought about ice. Synthesizing ideas that had largely been developed by others, Agassiz argued that erratic boulders and clay deposits could be explained by the action of a huge ice sheet that had once covered much of Europe and North America. The idea of an ice age raised many questions (had woolly mammoths really roamed the English countryside at the same time as humans?), but none more challenging than what it meant for the accepted history of the earth. The existence of enormous ice sheets in the past implied something that

was almost inconceivable to most people at the time: that the earth
might have been colder in the recent past.[9]

The reason this was such an unimaginable idea at the time was
thanks to the work not of geologists but of another type of investigator. These men did not typically spend much time mucking
about on the sides of mountains. Nor, when it came to it, did they
spend much time on the glaciers of the Alps. Brothers William and
James Thomson were among the most able of this brand of scientist. Armed not with stories but with mathematics, they made predictions that were best tested by experiments in the lab, where the
precision of their analysis could yield spectacular results. Unlike the
geologists, who saw history as a product of countless contingent
accidents, these men saw time as ahistorical and uniform, the kind
of time that unfolded in the belly of a steam engine. Their sacred
text was Newton's *Principia Mathematica*, and they hoped to do for
the earth's physics what Newton had done for celestial physics—to
provide equations that could perfectly account for its motions.

These men—today we would call them physicists, though the
term was only just starting to be used—were obsessed with energy,
and in particular with the transformation of heat into a form of
energy that could be harnessed as work. But though their working environments were far removed from the sites of coal mines
and railway cuttings, commerce played a role in their research as
surely as it did in the discoveries of the geologists. The industrial
revolution powered itself quite literally with the heat of the sun that
was stored up in the black deposits of coal. The heat of the sun and
the work it could do motivated the theoretical calculations these
men made and the experiments they did on all manner of things.
They studied the behavior of metal under pressure (a key factor in
designing boilers that did not explode), the way a certain amount of
work always generated a certain corresponding rise in temperature,
and how to design steam engines that were as efficient as possible.
This research led them to make predictions about how heat acted
not only in the workshop or the laboratory but in the crevices and
crannies of the earth itself. Deduced as they were from equations
that described the behavior of energy and matter, the conclusions

of men like the Thomsons furnished a kind of mathematical bedrock for the new sciences of the earth along with the increasingly consequential industries seeking ever more efficient ways to transform coal, via steam engines, into work.

One of the key ideas to emerge from this new way of thinking was that the universe, and everything in it, including the earth, was gradually and inexorably cooling down. This so-called heat death of the universe was a depressing fact of life, against which everything (with the important exception of God himself) seemed powerless. It followed that the earth's past must consist of a very steady, very boring, and very uniform cooling like that of a long-forgotten cup of coffee. Earth, in other words, could only have been warmer in the past, not colder. Recently unearthed fossil remains of coral reefs, seashells such as the (now tropical) pearly nautilus, and warm-weather plants such as palm trees and cycads that had been found in Northern Europe provided further, seemingly incontrovertible evidence that the earth had been warmer, not colder, in the recent past.[10] Taking into account the mathematical calculations of the physicists and the fossil evidence of warmer climates in the past, the cooling earth theory seemed an inescapable fact.

But the earth provided contradictory signs. Tumbling out of the crumbling walls of railway cuts and the dank tunnels of the coal mines came evidence that supported Agassiz's theory of the ice ages. How, then, to reconcile the seeming incompatibility between these different kinds of evidence about the earth's past? Some suggested that the continents themselves had been elevated in the past, allowing ice sheets to form at cooler altitudes and leaving the overall climate of the earth unchanged. But as the geological evidence accumulated for just how widespread ice sheets had been, this view became increasingly untenable. William Hopkins, a highly skilled mathematician who coached dozens of the best mathematicians of his generation as a tutor at Cambridge, calculated that it would be necessary to lift the entire continent of Europe above 10,000 meters to account for the ice age this way, a circumstance which, he added, "all geological experience assures us would be impossible without leaving numerous telltale signs which do not presently exist."[11] Nevertheless, the contradictory evidence of both a colder and warmer past on the planet demanded a new explanation. Hopkins had a mathematical answer to the dilemma. While it was true that the interior of the earth was cooling, Hopkins's calculations showed that it had already cooled to such an extent that the central or "primitive" heat of the earth (leftover from its fiery formation) was contributing a vanishingly small amount of heat to the earth's surface—just one-twentieth of a degree.[12] If the leftover heat contributed such a small proportion to the earth's surface temperature, the "problem" of heat death as a factor in terrestrial climate more or less disappeared. No longer did geologists have to find a way to link changes in the earth's climate at its surface to the slow cooling of its molten core. Hopkins confidently, but somewhat unhelpfully, declared: "We must manifestly seek for other causes to account for the changes of temperature which mark the more recent geological periods."[13] Whatever had caused the changes in the earth's climate that had occurred relatively recently, it was not the effect of the earth's cooling core. It remained to be seen what those other causes might be.

Hopkins had resolved the matter of the earth's cold recent past in relation to the overall cooling of its interior, a once insurmountable objection. By 1859, the year that Tyndall and his men braved the blizzard on the Mer de Glace, Agassiz's ice age theory was largely accepted, but two important questions remained unanswered. First, a theory was needed to explain how, precisely, the glaciers and ice sheets that Agassiz imagined covering so much of the Northern Hemisphere had moved. Second, and even more fundamentally, scientists still lacked an explanation for what had caused the earth's climate to cool so dramatically in the past. To answer the first of these questions, it was necessary to go, as Tyndall had, to the ice and study its motions. To answer the second would require traveling farther still, beyond the bounds of the earth and into the solar system itself.

In the late 1850s and early 1860s, a Scot named James Croll, an almost exact contemporary of Tyndall, was working quietly as a

caretaker at a small college and museum in Glasgow. Unlike Tyndall, who had, by his thirties, attained one of the few full-time pro-
fessorships in physics in Britain and with it a measure of personal
fame, Croll was unknown. He had not had any education to speak
of, but had discovered books as a young child anyway. Soon, a passion for natural philosophy emerged. This he nurtured even as he
worked a series of increasingly odd jobs, including the tending of a
tea shop for which the taciturn former carpenter was astonishingly
ill suited. For nearly three decades, he passed his time reading, independently and widely, in the process developing a taste for theoretical rather than empirical works. Facts alone held little interest for
him. He wanted the warp and woof of theory that held the facts—
the world—together. In the early 1860s, taking advantage of his light
work duties, he immersed himself in the study of what he called
"the then modern principle of the transformation and conservation
of energy and the dynamical theory of heat," reading the works of
Tyndall alongside those of Faraday, Joule, and William Thomson
on heat, electricity, and magnetism. At the same time, he followed
the unfolding developments in the debates over the "question of the
cause of the Glacial epoch."[14]

An autodidact with an idiosyncratic cast of mind, Croll was an
outsider to these debates, with neither institutional affiliation nor
professional training in any of the subjects. His remove gave him a
perspective—and a freedom—that enabled him to make his greatest cognitive leap. In 1864, he published a paper arguing that the
causes of changes in the earth's climate are to be found nowhere
on earth. Instead, Croll believed that it was in the subtly wobbly
dance of the earth around the sun that the cause of the ice ages was
to be found. The plural here is important—part of the reason Croll
reached outside the planet was to explain why he thought there was
not just one ice age in the past, but a series of alternating glacial
and warmer interglacial periods (evidence for alternating glacial
and interglacial periods had recently been uncovered by Archibald
Geikie in the form of layers of organic matter found in deposits of
glacial drift). Before Croll, eminent scientists had considered this
same astronomical possibility, including Alexander von Humboldt,

FIG. 2.6. James Croll, a self-taught theorist of climate, impressed Charles Darwin and John Tyndall with his argument that changes in astronomical cycles had produced multiple ice ages on the planet via "secondary causes" on Earth, such as variations in the reflection of sunlight by ice and cloud and corresponding shifts in winds and ocean currents.

Charles Lyell and, most influentially, the astronomer John Herschel. Herschel had noted that given the long-term gravitational perturbations affecting the earth, there would be moments, when its orbit was especially elliptical (or squashed), that winters would be longer and summers shorter. But Herschel quickly neutralized the implications of this fact in accounting for the ice ages by pointing out that the overall amount of sunlight hitting the earth would always be the same—the long winters would be compensated for by very hot summers.

Croll's innovation was twofold. First of all, he discounted more or less all the evidence of the geologists. He was blunt about his lack of interest in what he called the "facts and details" of science in favor of the more attractive (to him) basic "laws or principles" which underlay the empirical facts. (He remarked, positively, about a job he'd held as a geologist in the civil service that "really did not require much acquaintance with the science of geology" and therefore "relieved my mind from having to study a science for which I had no great liking, and thus allowed me to devote my whole leisure hours to those physical questions in which I was engaged.")[15] Croll was an unrepentant big-picture thinker. Having dispensed with the constricting assumptions of the geologists, confined to the lowering and rising of continents or the comings and goings of floodwaters and the pesky details of the scattered evidence, the second part of his innovation was to grasp at as big a cause as he could summon, the variable eccentricity of the earth's orbit. What he did next was his real leap. Rather than accepting Herschel's statement that changes in the earth's climate could not be explained by variations in its orbit, he suggested that what he called "secondary causes" operating not on the scale of planetary orbits, but back on the earth itself, were more than capable of accounting for the ice ages.

Water was the medium by which heat traveled around the globe, and the secondary causes to which Croll now turned were a function of the complex interplay between water in all its forms. These secondary causes were primarily the result of the physical properties of water. Even if the total sunlight in a year is held constant, cooler winters would mean more snow. As snow accumulated from year to year, its reflective qualities would come into play. Because snow is white, it reflects most of the light and heat from sunlight back to space, further cooling the earth. Here was a positive feedback mechanism (though Croll did not use the term) that could begin to account for the steady increase of snow-cover needed to generate an ice age. These cooler conditions begat further cold by increasing the tendency of cooling fogs to form over the snowpack, further insulating it from the heat of the sun. As the temperature gradient between the cold poles and the warm tropics increased, the trade winds would blow more toward the equator, deflecting the Gulf Stream to the north and its sister current, the South Equatorial Current, to the south, further increasing the heat imbalance.

The net result was a cooling planet that drifted into an ice age. And so it went, according to Croll's prodigious imagining, until the gravitational forces shifted and Earth's orbit assumed a rounder, less elliptical orbit, allowing the ocean currents to shift, snow to melt in the summer, and the feedback mechanisms to work in reverse, accelerating melting and warming where they had once accelerated cooling.

Croll's theory of climate change was based on global factors that were not geological in origin but physical. Croll imagined the movement of heat through the atmosphere and the oceans and the ice sheets as a means of explaining dramatic and sustained changes in climate and therefore had no need for the long, slow, and monumental rising and falling of continental-scale landmasses on which Lyell had staked his reputation. "The *cause* of secular changes of climate," he wrote, "is the deflection of ocean currents, owing to the physical consequences of a high degree of eccentricity in the earth's orbit."[16] Croll was unapologetically aligned with the physicists. He was not concerned by the lack of geological evidence for his theory. In fact, he claimed that the very absence of geological data was a form of evidence in favor of his theory. It was the nature of the ice ages that the erosive action of glaciers destroyed the evidence of their own (successive) passages. Croll was, like Tyndall, comfortable in drawing inferences based on fundamental physics. He was confident enough in his physics, and the logic of his inferential apparatus, to take his assumptions to their logical extremes. Scale was no object. If the logic led him to view the ice ages on this planet as a function of astronomical shifts married to global physical dynamics, then so be it.

In spite of his lack of standing in the scientific community, Croll had come up with a theory too compelling to be ignored. It piqued the interest—and raised the hackles—of some of the most prominent thinkers of the day. In the process of revising the tenth edition of his great work, *Principles of Geology*, Lyell wrote to ask his friend the great astronomer John Herschel what he thought of Croll's theory. Convinced of his own ideas about the gradual (and uniformly acting) changes that affected the earth's climate, Lyell couldn't completely disregard what he acknowledged was compelling evidence to the contrary. "I feel more than ever convinced that changes in the position of land and sea have been the principal causes of past variations in climate, but astronomical causes must of course have had their influence and the question is to what extent have they operated?"[17] That indeed was the question. To Lyell, climate change occurred primarily as a result of geographical change—the rising or falling of landmasses, the subsequent changes of sea level, and the blocking or opening up of ocean currents.[18] He considered neither astronomical nor the secondary physical mechanisms of warming ocean currents and coolly reflective ice to be sufficient to account for the changes in climate the earth had witnessed. Herschel's reply was not reassuring. Croll's astronomical causes were, according to the astronomer, "quite enough to account for any amount of glacier and coal fields." Given the right astronomical conditions, Herschel continued in grudging acceptance of Croll's theory, "any amount of glacier you want is at your disposal."[19]

As well as the reluctant acceptance of geologists such as Lyell, Croll found other allies who were more enthusiastic. Theory could offer a powerful way out of the conundrums caused by too much, and too complex, data. Reading the geological evidence that took the form of messy, complicated drift deposits had challenged scientists for decades. Drift was chaotic, disordered, and largely devoid of fossils. All the tools that geologists had so far developed to enable them to make sense of the structure of the earth depended on the presence of fossils (to allow comparative dating) and the assumption that sediments were gradually deposited over time and could thus serve as standardized indexes of past change. Drift obeyed none of these rules and offered none of these tools. Charles Darwin was forever embarrassed by his failure to see the evidence for glacial motion in the escarpments of Wales that he visited in 1831. Once he'd learned to read the landscape, it seemed clear to him that such features could only have been caused by massive ice sheets. Darwin was an enthusiastic reader of Croll's work, writing to tell him that "I have never, I think, in my life, been so deeply interested by any geological discussion. I now first begin to see what a million means, and I feel quite ashamed of myself at the silly way in which I have spoken of a million years. . . . How often I have speculated in vain on the origin of the valleys in the chalk platform round this place, but now all is clear."[20]

Less easy to see, even for those who were trained to look, was evidence of both advancing and retreating glaciers. In 1871, James Geikie (who had worked with Croll at the Geological Survey of Scotland since 1867) published a book outlining his own theory of the ice age—or, rather, ice ages, because Geikie's signature claim was that the ice age was actually a series of glacial periods punctuated by warmer interglacial periods. Thanks to Croll, Geikie was emboldened to look for the causes of terrestrial climate in the solar system. Dissatisfied with the idea that the rise and fall of great land-masses could explain such dramatic changes in climate as the fossil and plant evidence from the past suggested, Geikie wondered, "is it not possible that a solution of the problem may be found in the relations of our planet to the sun?"[21]

Geikie was a geologist by training and inclination, but he would not have arrived at his theory of alternating warm and cold periods on the basis of the complex and fragmentary geological record alone. Without Croll's big idea, Geikie would have had neither the confidence nor the insight to make his claim that there was not simply an ice age but, as one historian has called it, an "eventful" series of ice ages. When he published a series of seven papers on the topic, Geikie was careful to lead with the geological evidence for his ideas—glacial deposits found in Scandinavia, Switzerland, and North America—only referring to Croll's theory of climate in the latter papers. In doing so, he deliberately made it seem as if he was proceeding inductively, as geologists did, by building up a theory out of the many bits of evidence about glacial deposits. That way was safer, and more convincing to the geological community, than proceeding assuming the theory was correct and using it to make sense of the complex records of glacial deposits—a deductive form of logic.[22]

The need to bring other kinds of thinking to bear on the problem of the ice ages made many people uncomfortable. It was difficult, when different methods of science were brought together, to know what counted as evidence, or facts, anymore. How valuable were theories that couldn't be tested? Sometimes, as with the case of Croll, it could be very useful indeed to grab hold of an idea, such as alternating glacial and interglacial periods, that could be used to sort a planet's worth of evidence. There would always be anomalies. The trick was to determine when, if ever, the anomalous data was heavy enough to bring the whole theoretical structure tumbling down. In the meantime, if the structure held and enough data fit, it seemed reasonable to ignore the few pieces that didn't.

Geikie was sanguine about the prospect of what we would call more interdisciplinary approaches to the big questions, with that of the ice age being most important. "As the circle of knowledge widens," he wrote in his 1874 book *The Great Ice Age*, "boundary divisions become more and more difficult to determine. Perhaps of no physical science is this more true than that of geology. At one time the investigator into the past history of our globe had the field almost entirely to himself, and the limits of his study were as sharply defined as if they had been staked off and measured. Now, however, it would be hard to say on which of the territories of his scientific neighbors he must trespass most. He cannot proceed far in any direction without coming in contact with some worker from adjacent fields. His studies are constantly overlapping those of the sister sciences, just as these in turn overlap his." This disciplinary crowding was, Geikie thought, itself proof of the ways in which all natural phenomena were knit together into a whole cloth. "It will, therefore, be a further proof of the unity of Nature, if those intricate problems which have hitherto baffled the geologist should eventually be solved by the researches of astronomers and the conclusions of physicists."[23]

Among those who found much to agree with, and be stimulated by, in Croll's theory was none other than John Tyndall. Given their shared commitment to physical reasoning, it's little surprise they gave each other mutual encouragement. The two men corresponded, with the more conventionally successful Anglo-Irishman offering encouragement to the unknown Scot. Croll had come to his theory of global climate change directly after completing work on the behavior of heat in solids. Like Tyndall, his understanding of the forces at work on planetary scales was underpinned by an understanding of the physics of the molecular. Also like Tyndall, Croll was no great mathematician. Both relied on a rather astonishing intuitive feeling for the way physical forces interacted rather than a mastery of complex mathematics. Tyndall praised Croll's use of a metaphor to describe the action of molecules: "your letter was interesting to me as an illustration of power to seize a definite physical image—the molecules acting as hammers was capital."[24] Tyndall embraced him as an intellectual fellow-traveler, a thinker who used images to make his points (and perhaps even to think with) and was unafraid to generate grand theories. "It is both amusing and interesting to me," wrote Tyndall in another letter, on studies of heat, "to trace the parallelism which has run between your thoughts and mine on the subject."[25]

While Croll remained resolutely fixed on the largest of scales—that of the entire planet—Tyndall made a career of moving between scales. He observed and generated theories about the very small, such as ice crystals and water molecules, and the very large, such as glaciers and mountains. Tyndall saw connections everywhere, but water in particular provided access to this central mystery, and beauty, of the universe: its continuity. It is strange but true, wrote Tyndall, that the "cold ice of the Alps has its origin in the heat of the sun."[26] He went on: "You cannot study a snowflake profoundly without being led back by it step by step to the constitution of the sun. It is thus throughout Nature. All its parts are interdependent, and the study of any one part *completely* would really involve the study of all."[27] For Tyndall, fundamental forces were never far removed from the largest scales of all. The tiniest changes that might occur in the deepest part of a glacier were connected not only to the entire glacier and its motions, but to the general physical properties at work everywhere on Earth and in the universe as a whole. Tyndall's imagination did not stop at any point, but continued ever upward and outward, linking earthly physics to the physics of the sun and the cosmos in a chain of physical connection.

The idea that Nature was continuous—that energy and matter were linked in an uninterrupted chain of events—was for Tyndall a kind of secular religion. While he expressed his conviction in this continuity more loudly—and insistently—than many of his contemporaries, he was not alone in seeking to use the physics of the building blocks of matter to explain and understand the most complex, large-scale phenomena possible. Where previously, those who studied the earth had been content to try to map and describe what lay before them—as naturalists and as geologists—in the middle decades of the nineteenth century, it seemed increasingly possible to discover the mechanisms by which matter shaped the earth, to explain rather than simply describe terrestrial phenomena. Glaciers, then, were the perfect laboratory not only for testing ideas about the history of the earth but for helping transform the sciences of the earth. "No branch of study will place us in closer connection with the workings of nature," wrote a reviewer on glacier writings by Tyndall, Agassiz, and Forbes, "or in a better position to observe how the most delicate physical elements combine to produce the most stupendous results, than that which concerns those vast masses of ice, the glaciers."[28] Glaciers helped reveal the internal actions of nature, to show how something as delicate as a crystal of ice could, when given enough time and alongside enough other crystals, change the shape of entire mountain ranges, and even entire continents.

Where Tyndall's special skill lay was in showing how the discoveries yielded by the mountains could be extended by experiments done back in the laboratory. In the Alps, Tyndall measured the motion of ice flows that filled entire mountain basins. Back in London he could continue his studies of the motion of ice at vastly reduced scales. While Tyndall was motivated by a pointed desire to "destroy" the theories of those he saw as his rivals, it was in the to-and-fro between mountains and laboratories—rather than in the battle for the definitive theory of glacier motion—that he would make his greatest contribution.

His laboratory was located in a convenient spot in the basement of the Royal Institution, on Albemarle Street just off Piccadilly, where he gave public lectures in his role as professor of natural philosophy. In the summer of 1856, immediately following his first excursion to the mountains with Thomas Huxley, Tyndall had returned to his laboratory and started fooling around with ice, turning impervious river ice into fissured glacier ice, making the telltale blue bands he had witnessed on the glacier appear, forcing cleavage into transparent ice of almost crystalline perfection.

The experiments were beautiful in their simplicity. He had a laboratory assistant make a series of hard wood molds. He used them to shape and squeeze ice, to try to imitate the action of the glaciers. What he wanted to show was that ice moves through a stutter-step of freezing and melting that happens at such minute scales of both time and size that it is indistinguishable, in its results, from the flow of water. While ice *seemed* to act like a viscous liquid—like treacle or honey—according to Tyndall it really acted like a brittle substance. Under the cover of thousands of tons of glacier, the ice melted and refroze in tiny but discrete increments. Tyndall used an ugly word, originally his mentor Michael Faraday's, to describe the process. Regelation. A series of alternating physical states, solid to liquid to solid, that occurred far beneath the surface, at the rough contact between the glacier and the earth. The special addition of pressure was a new twist provided by James and William Thomson. Together, the Thomsons predicted and then showed by experiment that pressure lowers the melting point of ice.[29] This means that at the bottom of the glacier, where the pressure of untold tons of ice and snow lying above is greatest, the ice will melt. Meltwater will flow away from the glacier and, having thus lowered the pressure incrementally, the base of the glacier will refreeze momentarily until the pressure increases sufficiently to melt it again. And so the cycle will continue.

The hum of atoms that Tyndall believed enlivened every bit of the universe, seen and imagined, rose to a pitch beneath the ice. Far below the surface, the ice hovered, almost tenderly, at its melting point. Across that tender borderline, the ice fractured and healed itself repeatedly. It broke under its own pressure and mended under that same pressure, releasing heat in the form of water. What seemed

fluvial was massive, skidding, a giant locomotive making its judder-
ing way down to the valley.

On January 15, 1857, Tyndall made the first of many presenta-
tions on his glacial research. In addition to presenting the theory
he and his friend Thomas Huxley had developed together, Tyndall
took the opportunity to attack the other leading theory of glacial
motion of the day, that of James David Forbes. Forbes was older
than Tyndall and had first traveled to the Alps, with Agassiz himself,
fifteen years before Tyndall had. Forbes had gained the enmity of
Agassiz after publishing a paper that the latter felt failed to credit
him sufficiently. Undaunted, Forbes continued to travel to the Alps
and publish papers based on the notion that the ice was, in fact, a
viscous substance similar to treacle. Tyndall and Huxley attacked
Forbes's sloppy use of the term *viscous*. Tyndall asserted that when
stretched enough, the ice would eventually turn brittle and snap.
The viscosity that Forbes diagnosed was only apparent, not real,
argued Tyndall.

Tyndall's theory does not sound so very different, to modern
ears, from that of Forbes. Both were convinced that ice flows like a
liquid; they differed only in the details of how it flowed. The difference between them lay in the kinds of evidence they used to make
their claims. For Forbes, the movement of glaciers was a matter of
geology—of understanding the forces shaping the earth at the largest scales. The micro-physics of how, precisely, the ice flows was
unimportant. For Tyndall, the matter was one that must be solved by
physical reasoning. Drawing on work by Hopkins and the Thomson
brothers, Tyndall suggested that the glacier was moving in increments. He admitted that these were tiny increments, consisting of
miniscule amounts of ice that melted and refroze at the very depths
of the glacier, but the point to be made was much larger. Ideas about
molecules and energy that had been gleaned from mathematics and
the most fundamental physical descriptions could be used to predict and to understand the behavior of the largest, messiest, and
seemingly most inscrutable of things, the glaciers of the Alps. In this
sense, the battlefield over which Tyndall and Forbes were tussling
was very large indeed. Not simply a matter of semantics, it was a

battle over the right to claim that a certain way of knowing the earth and a certain kind of explanation was more truthful than another.

Not everyone agreed with Tyndall's version of science. Before agreement could be reached about the nature of the ice ages, the history of the earth, or the movement of glaciers, agreement had to be achieved on what an answer—a theory—might look like. Tyndall had tried to win the battle by linking his heroic exploits in the mountains to his disciplined experiments in the laboratory. In both cases, he suggested that he alone (despite the near-ubiquitous presence and contribution of assistants and porters) was capable of revealing the hidden truth of phenomena that seemed one way but really were another. In the case of the ice, the seeming flow of the glaciers was actually incremental regelation.

William Hopkins considered that both Tyndall and Forbes were each a bit right and each a bit wrong in their proposition of something they both called a "theory of glacial motion." Too many of the "numerous discussions which have taken place during the last twenty years," wrote Hopkins, were only partial and incomplete theories. What was lacking was a "complete and sufficient theory founded on well-defined hypotheses and unequivocal definitions, together with a careful comparison of the results of accurate theoretical investigation with those of direct observation." What Hopkins wanted to call a theory was something that looked like physics— based on a well-defined hypothesis and absolute definitions—that also explained the phenomenon that had been observed by the geologists. As far as Hopkins was concerned, both Tyndall and Forbes were guilty of calling a total theory what was merely partial description of one of the ways a glacier could move. "The Expansion Theory ignored the Sliding Theory, though they were capable of being combined," wrote Hopkins, "the latter theory was equally ignored by the Viscous Theory . . . [and] the Regelation Theory is not properly a theory of the motion of glaciers, but a beautiful demonstration of a property of ice, entirely new to us, on which certain peculiarities of the motions of glaciers depend." The best and final theory, pointed out Hopkins, would be one that required no "qualifying epithet" to distinguish it from a rival claim.[30]

In making the move from the Mer de Glace to the laboratory, Tyndall was trying his best to come up with such a final theory. He was trying to forge a link between the geologists, such as Forbes, whose identities were bound up with muddy boots, sturdy instruments, and physically taxing expeditions to mountaintops and glacial fields, and the physicists, such as William Thomson and William Hopkins, who appealed to physics and mathematics to explain natural phenomena. Much of what Tyndall was doing in the Alps and in the laboratory anticipated later developments in the earth sciences that brought together mathematical physics and descriptive geological approaches.[31] It would be wrong, though, to see Tyndall as fully modern. He was a qualitative rather than quantitative physicist. Analogy and metaphor were his tools, not mathematics. His most significant achievement was to link different ways of knowing. He brought together the experience of being on the glacier—the awe, the terror, and the particular insight into natural phenomena that it made possible—with the experimental investigations he did in his laboratory, a place from which awe, terror, and the specificity of landscape was expressly excluded. By doing both things—and calling both things "science"—he was staking a claim to what science might be.[32]

He performed a similar trick in his writing, drawing distinctions between different ways of knowing even as he brought the two together. His 1860 book *The Glaciers of the Alps* was divided cleanly into two parts.[33] The first, which Tyndall called narrative, contained headings such as "Expedition of 1856" and "First Ascent of Mont Blanc, 1857." The second, which he called science, contained such chapters as "light and heat," "origin of glaciers," and "the colour of water and ice." Intimately related, the two were also best kept separate. "The mind once interested in the one," Tyndall cautioned, "cannot with satisfaction pass abruptly to the other."[34] He knew almost instinctively how to keep an audience member or a reader engaged. "Once upon a steep hard slope Bennen's footing gave way," began one episode about an adventure he and his Swiss guide had on a descent from the summit of the Finsteraarhorn, "he fell, and went down rapidly, pulling me after him. I fell also, but

PEAKS, PASSES, AND GLACIERS.

A Series of

EXCURSIONS BY MEMBERS OF THE ALPINE CLUB.

EDITED BY JOHN BALL, M.R.I.A. F.L.S.

PRESIDENT OF THE ALPINE CLUB.

"Per Nives acumptuenas Rupesque tremendas"

LONDON

LONGMAN, BROWN, GREEN, LONGMANS, AND ROBERTS.

1859

FIG. 2.7. The public enjoyed stories of daring exploits by the members of the Alpine Club, established just two years before this publication. From the title page of *Peaks, Passes, and Glaciers, A Series of Excursions by Members of the Alpine Club* (London: Longman, Brown, 1859).

turning quickly, drove the spike of my hatchet into the ice, got good anchorage and held both fast."[35] This was material designed to keep the attention of a young boy or man—Tyndall's imagined (and it seems preferred) audience. Having captured it (and impressed the reader with his own stamina and bravery), Tyndall hoped to carry his readers with him into the more austere topics such as the curious veined structure of the glaciers or, more to the point, the mechanism responsible for their motion.

The Glaciers of the Alps was a very popular book, selling many copies and bringing Tyndall firmly into the center of a fashionable circle of London intelligentsia. The public seemed to clamor more for stories of danger and heroism than they did of glacier motion, but on the whole, the idea that the search for knowledge motivated the risky feats made them seem more rather than less heroic. Just as John Franklin's expedition in search of the fabled Northwest Passage (and those sent out subsequently to search for the lost ships) mixed national pride with the nobility of scientific exploration, so Tyndall's work blended vicarious thrills with the cooler appeal of participating in the increase of scientific knowledge.

Tyndall's scientific style wasn't to everyone's taste. A group of scientists found common cause in opposing him. He was, they thought, the worst possible combination: a dangerously unchristian show-off seriously lacking in mathematical skills. James Clerk Maxwell used the full force of his own literary sensibility to attack the Anglo-Irishman, coining a term, "Tyndallize," to refer to the theatrical manner in which Tyndall communicated. In an 1863 manuscript poem that was shared privately among the group of Tyndall skeptics, an anonymous author (almost certainly Maxwell) held little back:

> There on a platform stood the fiend
> His wide mouth grimaced in smile
> Upon his right, the electric light
> And on his left a [galvanic] Pile
>
> And lo! The well dressed multitudes
> Pressed forward in profusions
> While scientific beggars sat
> On the door of the Institution.[36]

In Maxwell's biting poem, Tyndall becomes a grotesque caricature, as do his fashionable audiences and all those excluded from the spectacle. As if the point weren't already clear, in his review of an essay by Tyndall in a popular magazine defending his theory of glacier motion, P. G. Tait commented that "Dr. Tyndall has, in fact, martyred his scientific authority by deservedly winning distinction in the popular field."[37]

These sorts of exchanges make it clear that though Tyndall may have won the battle for the public's attention, he had done so at the cost of alienating some of his scientific peers. His success at lecturing was not enough to win him victory in the matter of glacier motion. But nor, for that matter, did Forbes. The dispute between the two men over how glaciers moved was never resolved. Petty concerns about priority and citation mounted, and the whole debate degenerated into little more than name-calling.[38] This was partly a result of the strong personalities involved, but it was also a measure of how much the debate was not about any theory in particular but, rather, about what counted as a theory. When the very boundaries of a problem are up for consideration, it is difficult, if not impossible, to recognize any given explanation as more complete than another.

Tyndall was dissatisfied with the stalemate the glacier debate had reached, but he was a restless thinker and, above all, a *doer*. He soon found a new project into which he could sink his prodigious energies and which would again unite his passion for the grandeur of nature and the precision of laboratory experimentation. He considered the new project, quite naturally, to be continuous with the old one. He was still fundamentally interested in the role played by heat in matters both of basic physics and in the complicated, messy reality of the earth itself. His trips to the Alps made to study the glaciers and how they moved started him thinking about gases, heat, and the sun's rays. It was impossible for him to be there, in the mountains, without thinking about the endless transfer of energy from one substance to another. In his own words, his work on glaciers had "directed my attention in a special manner to the transmission of solar and terrestrial heat through the atmosphere."[39] His new project would be an experimental investigation of how heat affects not solid objects, such as ice, but gases, including those found in the earth's atmosphere. It is this work for which he has now regained a measure of remembrance, as an early discoverer, along with Joseph Fourier and Svante Arrhenius, of what we today call the greenhouse effect.

He spent time in early 1859 in the basement of the Royal Institution with a specific question in mind: How much heat could different gases absorb? To answer this question, Tyndall created a complete and controlled artificial environment, part electrical apparatus, part cloud chamber. It never got a name of its own. It was a complicated thing, comprising a sealed glass tube within which he could release gases of different types, a steady source of artificial heat (provided by a gas flame and a cube of boiling water), and a precise method for measuring the absorption of that heat by the gas, via an instrument (only recently invented) called a galvanometer, which would measure the difference between the current transmitted through the tube with or without the gas inside it.[40]

In concept, simple enough (perhaps). In fact, the device spun off problems like a Catherine wheel. Even when no charge was present, the galvanometer's needle turned of its own accord, as much as thirty degrees from neutral. After much tinkering, Tyndall finally figured out that the copper used to make the wire of the coil was tainted by magnetic metals. A purer, less magnetic copper reduced the deviation from thirty degrees to three. But it was still not good enough. The absorptive properties Tyndall hoped to measure might be very small indeed. Three degrees of error in the instrument could obscure any effect he might be trying to measure. He finally thought to unwrap the green silk that covered the copper wire. Some compound containing iron had been used to dye the silk green. The bare wire, wrapped in white silk with clean hands, gave no deflection of the needle.

At first, despite the improvement in the apparatus, he saw nothing in the gases that he studied. Contriving a constant source of heat was a major challenge. He spent weeks in the spring of 1859 trying to get a result, during which time he often despaired. "The course of the inquiry during this whole period was an incessant struggle with experimental difficulties." It was as different as could be from the moments of instant epiphany he experienced in the mountains,

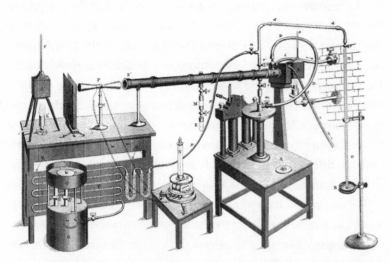

FIG. 2.8. The apparatus John Tyndall used to measure the absorption of heat by different gases, including water vapor, which could explain "all the mutations of climate which the researches of geologists reveal." John Tyndall, "The Bakerian Lecture: On the Absorption and Radiation of Heat by Gases and Vapours, and on the Physical Connexion of Radiation, Absorption and Conduction," *Philosophical Transactions of the Royal Society* 151 (1861): 36.

where all was revealed to him as abruptly as the shifting of a cloud from the face of the sun.[41] And then, on May 18, 1859, after nearly two months of constant work with the device, he had a breakthrough: "Experimented all day; the subject is completely in my hands!" The next day he continued, "Experimenting, chiefly with vapours, coal gas wonderful—ether vapour still more so."[42]

But then, seemingly unaccountably, Tyndall broke off his experimental labors, returning to the Alps to make more glacier studies. Though he was not employed at an academic institution, he kept academic timetables—lecturing and experimenting in the fall and spring, with summers off. The break he took in June 1859 was a recurrent and natural one for Tyndall. Summers were devoted to the Alps, and to the Alps he went in the summer of 1859. It was only in September 1860 that he would return to the apparatus, spending seven weeks fine-tuning it, trying and rejecting new sources of heat. He experimented up to ten hours a day. Over the course of the next seven weeks, he worked nonstop, experimenting from eight to ten hours a day. He studied sulfuric ether, ozone, olefiant gas, carbon bisulfide, ethyl iodide, methyl iodide. The list went on, stretching to dozens of substances. By late October, he had almost run his way through the long list of elements he'd set as his primary course of work. Slowly, he'd learned how to still the motley assortment of molecules in the room sufficiently so that the miniscule effects he sought were sensible to the instrument. For it had been a matter of increasing frustration that the simple substances he investigated were almost to a fault extremely poor absorbers of the heat that radiated from the boiling cube. They varied, it was true, and Tyndall worked hard to capture the silent music in these jittery numbers.[43] It was still not good enough. This work, too, was ultimately unsuccessful, and he rejected all of his findings. It was a trying time: "a continued struggle against the difficulties of the subject and the defects of the locality in which the inquiry was conducted."[44]

One problem was getting a steady heat source. In November 1860 he had better luck. The air of the laboratory, freed from its moisture and carbonic acid, produced a deflection of about one degree. Oxygen obtained from chlorate of potash and peroxide of manganese produced the same deflection, as did nitrogen, hydrogen produced from zinc and sulfuric acid, and hydrogen obtained from the electrolysis of water. He worked especially hard to obtain a pure specimen of oxygen, first obtaining a sample from electrolysis, and then sending it through a series of eight bulbs containing a strong solution of iodide of potassium, depriving the oxygen of its ozone. This too produced a deflection of just one degree. Then he tried the oxygen that had not been passed through the successive baths of iodide of potassium and still contained its complement of ozone. The needle jumped to four degrees. What this meant was that ozone was three times more potent an absorber of radiant heat than oxygen alone.

On November 20, something even more surprising happened. He first measured the absorption of heat produced by air that had been rid of its moisture and its carbonic acid. This was a negligible amount. That was unsurprising, given the readings he'd been gathering for other elements. But then came the unexpected result. Air that had been taken direct from the laboratory deflected the needle an incredible fifteen times more. Tyndall subtracted the effects of the carbonic acid from this and was still left with an amazing result. The invisible moisture carried by undried air was responsible for blocking thirteen times more heat than oxygen alone.

After a total of fourteen weeks of experimentation, Tyndall was able to report results in his Bakerian lecture of 1861. He saved his biggest discovery for the tail end of the paper. After describing the small deflections produced by substances such as chloroform and alcohol, he came to a point of "considerable interest" having to do with the relationship between the atmosphere and what he called solar and terrestrial (i.e., earthly) heat. The curious result was this: Air that had been rid of all moisture and other constituents absorbed very little heat, while air that had been taken directly from the laboratory produced an absorption up to fifteen times greater.

Even very small alterations in the amount of the key gases—water vapor, carbon dioxide, and hydrocarbon vapors—could change the amount of heat trapped by the atmosphere dramatically, thereby warming the planet. This was a mechanism that could, potentially, explain both the ice ages and the warmer periods whose existence was suggested by the fossil record. It explained why mountaintops were so cold, even though they were closer to the sun, and why the sun was hotter at midday than evening. The key to it all was the split in the nature of aqueous vapor. Though water vapor exercised a "destructive action," to use Tyndall's phrase, on the rays of radiant heat emitted by a cooling earth, it was completely transparent to light rays. This made all the difference. The light that reached Earth from the sun passed through the aqueous vapor easily and was absorbed by the earth, which then radiated heat back outward, as any rock warmed by the sun will do. That heat was then trapped by the aqueous vapor, which acted like a great blanket, swaddling the earth in heat which would otherwise be lost to space. Variations in the amount of aqueous vapor, Tyndall surmised, could account for many, if not all, of the changes in climate that fossil records and geological strata revealed. No longer would it be necessary to theorize changes in the density or height of the atmosphere or in the elevation of entire continents to explain the different degrees of heat reaching the earth. Instead, "a slight change" in the amount of water vapor in the atmosphere was sufficient to produce "all the mutations of climate which the researches of geologists reveal."[45] He needed to repeat the experiment in other locations, with other samples of atmospheric air, to eliminate any possible interference caused by dust or other particles in the air. But the implication was startling. "It is exceedingly probable," wrote Tyndall, "that the absorption of the solar rays by the atmosphere . . . is mainly due to the watery vapour contained in the air."[46]

Tyndall's patient work in a quiet basement in London had produced results that could explain how changes in the earth happened at the largest scales imaginable—in terms of both time and space. The absorption of heat by water vapor in the atmosphere clearly affected the entire climate of the globe. He was not shy about saying so in both the lecture he gave and the paper he published summarizing his results. His paper was read aloud to the fellows of the Royal Society and was chosen to be the Bakerian lecture for the year—a special honor.

He didn't have much time to enjoy his success before an unwelcome letter came from a German physicist named Heinrich Gustav Magnus, asserting priority. Tyndall was ready for it. He'd placed a preliminary notice before the Royal Society back in May 1859 for just such an eventuality. The notice didn't contain all of his subsequent results, but enough to stake a claim. But there was more serious trouble. Magnus had come up with results that were different from Tyndall's—diametrically so—on the matter of water vapor. Magnus had found that dried air absorbed more (only slightly, but still more) heat than moist air.[47]

Tyndall's response was hard work. "Self-chastening," he went so far as to call it, deliberately calling up the religious comparison. The research had already been an "immense and arduous toil," but that was as nothing to the next period of work, spurred by the presence of an unwelcome competitor.

He spent every weekday for the next four months proving Magnus wrong. With every improvement in his technique, Tyndall could see the effects of aqueous vapor more clearly. For moist air, he obtained deflections of the galvanometer's needle of forty-eight or even fifty degrees, while the dried air shifted the needle by only one degree.

To his relief, the more he became a master of his equipment, the greater the difference between the absorptions. He developed new and elaborate methods for drying the air. A massive block of glass was ground to dust in a mortar, boiled in nitric acid, washed with distilled water, and finally dried before being moistened by pure sulfuric acid. These fragments were then introduced into a U-shaped tube to prevent any contact between the sulfuric acid and the cork which stoppered the tube and which could otherwise undo the good effects of the drying process previously achieved. Pure white marble was ground for use with the caustic potash. Tyndall prepared new drying tubes daily to ensure they were equally effective. With these tubes, Tyndall had invented a way to rid the air of carbonic acid and moisture independently. And with his galvanometer and tube, he could measure the different absorptive capacities of various molecules.

Magnus remained unconvinced, questioning whether the findings would hold for air from locations other than Tyndall's laboratory. Tyndall took up the challenge. He had plenty of friends who were only too happy to help. Very quickly, he managed to obtain samples of air from places that were much clearer in atmosphere than the very center of London, if not as pellucid as the Swiss skies. His friends sent him air from Hyde Park, Primrose Hill, Hampstead Heath, and the Epson racecourse. He also received samples from two locations on the Isle of Wight, one a beach near Blackgang Chine. Consistently, Tyndall found that these samples all absorbed sixty to seventy times more heat with their water vapor than without it. Once again, Tyndall had managed to show that findings determined in one location—be it the Alps, his Albemarle laboratory, or Hampstead Health—could be duplicated elsewhere.

Having taken on the physicists, Tyndall was bold with the meteorologists. He was unafraid to direct them to his results, which offered what he termed "absolute certainty." Tyndall stated categorically that "the withdrawal of the sun from any region over which the atmosphere is dry, must be followed by quick refrigeration." This was, Tyndall freely admitted (even bragged), "simply an a priori conclusion." Such were the fruits of laboratory experimentation. He was confident enough to state that be believed no meteorological experience would contradict it. Once gained in the laboratory, truths could not be lost. He boldly asserted that ten percent of the "entire terrestrial radiation is absorbed by the aqueous vapour which exists within ten feet of the earth's surface."

His conclusions about aqueous vapor accounted for all sorts of climatological findings. It explained that the great twist of cloud that commonly mimics the course of the great rivers, the Nile and the Ganges, owes its existence to the "chilling of the saturated air above the river by radiation from its vapour." It accounted for the huge differences in temperature at high altitudes between the air, which often remained very cool, and the ground, which could warm up readily in the sunshine, which his friend Hooker had noticed in the Himalayas.[48] The effect held in Europe. A descent from Mont Blanc, hip-deep in snow, had been suffered in blazing sunlight and almost unbearable heat, despite the snow all around. It also explained how, in the arid regions of central Australia, temperature fluctuations of forty degrees were quite common, more than double the range found in damp London. Correspondingly, it explained the intense daytime heat of the Sahara and its nighttime cold.

As impressive as it was to be able to provide what Tyndall considered conclusive proof about the causes of the earth's climate, this was not, and never had been, Tyndall's primary goal. As with his studies of the glacier, Tyndall was really interested in molecular physics, with how heat affected molecules in their most basic form. When he described the alteration of the climate, he said that it revealed the "effect of our atmosphere on solar and terrestrial heat." Heat was always his primary focus, and the atmosphere was of interest primarily in its role as an interruption in the journey of heat. The work he did on gases was important to Tyndall because it revealed a "purer case of molecular action" than had ever been

FIG. 2.9. John Tyndall in 1877. He struggled throughout his life to reconcile his desire for solitude and for fame.

quickness of movement and thought made him a captivating lecturer and a graceful climber, but it redounded against him in the sphere of human emotion. His lifelong tendency to conflict led to disputes for which he is now most known—with the religious establishment on the efficacy of prayer or the relationship between cosmology and religion, as well as over a range of topics including spontaneous generation, the conservation of energy, and, as we have seen, glacier motion. More personally, and ultimately more consequentially, his inability to find a way to regulate his mental dynamo, to soothe his mind and gentle into sleep, led to a lifetime of trouble. Insomnia plagued him, and he turned increasingly to medication to help him sleep. He recorded anguished descriptions of sleepless nights when he sought relief in the chloral his doctor prescribed him. It sometimes took two or three doses before he finally fell into an artificial rest. One otherwise unremarkable day, his wife Louisa made a fatal mistake, administering an accidental extra dose of chloral. They both knew instantly what it meant. "Louisa, you have killed your John," Tyndall is reported to have said. The doctor was sent for, and every attempt, including shocking him with a galvanic pile, was made to revive the increasingly unresponsive scientist. It was all to no avail, and Tyndall died in the evening of December 4, 1893, age seventy-two.

Louisa, just forty-seven at the time, never forgave herself. She devoted the remainder of her life, some forty-seven additional years, to the task of writing his biography. Weighed down by guilt and the scope and scale of the material, she proved unable to complete the task. When she died in 1940, Tyndall's own generation was long gone. Few remembered Tyndall. The papers and notes relating to Louisa's mammoth project were eventually turned over to a pair of scientists who agreed to finish the task. Unlike his peers, whose descendants or students had completed the typical "Lives and Letters" or "Collected Papers" soon after their deaths, Tyndall—having had no students and only a guilt-stricken widow—received no such encomia. The long-delayed biography appeared in 1945 in a nation preoccupied with recovery from World War II, which had largely forgotten him and which accordingly took little notice.[49]

Today, Tyndall is receiving his measure of remembrance. His findings now look prophetic, the first glimmerings of understanding about the way our global climate works and the way human beings have, unintentionally, altered that climate in dramatic ways. Looking closer, it is clear how very different the preoccupations of Tyndall and his contemporaries were from ours. That his obsessions— ice, glaciers, water vapor, heat—look so much like ours do today should not fool us. Tyndall's study of heat was grounded in his deep appreciation of the recently revealed laws of thermodynamics, not in an appreciation of the living, green earth to which we are now

FIG. 2.10. John and Louisa Tyndall in the library at their home, c. 1887.

so attuned. It is not too much to say that the second law of thermodynamics, with its commanding prescription about the very cold
future of the cosmos and all its contents, gave him the eyes with
which he saw the world.

Today we are intensely aware of the connectedness of the parts
of the earth, but our version of what it means to be linked to each
other and to physical phenomena is very different indeed from
Tyndall's. We no longer look at the world through entropy-tinted
glasses. Nor do we find ourselves struck by the unimaginably long
time spans that have gone into making our planet—and which
stretch out ahead in the future of the universe. Instead, we see vola-

tility and ever-diminishing quantities of what was once a seemingly endless resource: time itself. Time, we feel, is running out as we rush to understand the mechanisms of our planet's climate and, just possibly, to make changes to our behavior to prevent an ever-hotter future.

The easiest difference to spot—and it is a glaring and significant one—is the fact that neither Tyndall nor his contemporaries imagined that human beings might pour so much of that other great absorber of radiant heat, carbon dioxide, into the atmosphere that it would change the climate of the earth. For all their imaginative musings about the past climate of the earth, and therefore the possible future climates of the earth, not one of them imagined that they had already embarked on the largest, and most momentous, experiment in the sciences of the earth ever undertaken.

Tyndall never lost his sense of wonder at the action of Nature, a wonder that was imbued with a feeling of awe at the almost infinite grandeur of natural beauty, the way it pervaded every molecule of creation with an almost uncanny reach. Even places to which humans had scant access—places where humans were not, in the normal course of events, expected to be—were drizzled with beauty. Despite (or is it because of?) his commitment to a purely materialist vision of the cosmos—one from which God had been excused as the origin of all causes—Tyndall was persistently moved to his very core by his own ineffable experience of wonder. Having rejected the idea that wonder could be a gift from God, he was left to wonder more deeply still at how such feelings could arise from the movement of energy through matter, from molecules in motion and nothing more.

He wrote feelingly of the remarkable power of the human imagination to peer behind the veil with which Nature obscured herself, but he never lost sight of Nature's deeper power. Entering the hut in which he and his companions spent the night during their 1859 expedition to the Mer de Glace, he noticed one more instance in which Nature outraced the faculties of men. Though the hut had been carefully closed up, fine snow crystals had found their way in through tiny crevices to form on one of the windows a "festooned curtain formed entirely of minute ice crystals. It appeared to be as fine as muslin; the ease of its curves and the depths of its folds being such as could not be excelled by the intentional arrangement of ordinary gauze."[50] Nature, without intention, produced beauty that dwarfed the greatest achievements of man. Explain *that*, Tyndall challenged his reader—and himself—time and time again.

Holding his hand to the windowpane and melting the ice beneath it, Tyndall watched it refreeze before his eyes as "atom closed with atom, and the motion ran in living lines through the pellicle, until finally the entire film presented the beauty and delicacy of an organism. The connexion between such objects and what we are accustomed to call the feelings may not be manifest, but it is nevertheless true that, besides appealing to the pure intellect of man, these exquisite productions can also gladden his heart and moisten his eyes."[51] Though it already felt strange to say it ("the connexion . . . may not be manifest"), Tyndall felt compelled to point out the link between emotions and the apprehension of order in nature. Emotions coursed through Tyndall in the same way that the heat from his hand coursed through the icy pane of glass. In both cases, the same physical principles were at work, and yet it was impossible to escape the uncanniness of a world—the world which Tyndall could not help but inhabit—where emotions and atoms were similar kinds. How could Nature make him feel so many things when feelings were nothing more than molecules? The thought provoked its own strange compulsion, a mental loose tooth to be explored again and again.

Throughout his life Tyndall experienced a symphony of emotion that was only heightened, not diminished, by his awareness of the mysterious materiality of his feelings. In Tyndall's appreciation of the beauty of Nature, there was always also a piquant awareness of this strange and exquisite paradox. The forms of Nature produced by physical laws could produce such intense emotion in human beings, who were themselves mere matter organized according to the same physical laws. There was, in the end, nothing to do but wonder, and keep looking. "In the application of her own principles," Tyndall wrote feelingly, "Nature often transcends the human imagination; her acts are bolder than our predictions. It is thus with the motions of glaciers; it was thus at the Montanvert on the day now referred to."[52]

3

SEE-THROUGH CLOUDS

The peak of the island of Tenerife appeared for only a moment as the RMS *Titania* eased into harbor, but Charles Piazzi Smyth was ready for it. He caught the clouds "unveiling for a moment the chief glory of the island, showing it for an instant as a reward after the toil of the voyage." He knew that his next vision of the peak would come only following a hard climb up the mountain. For the time being, he exulted in the chance to see "a higher and purer sphere."[1]

Though it felt serendipitous, the vision of the peak was far from an accident of cloud and wind. Instead, it was a direct consequence of what Piazzi Smyth called a "certain," as in definite, "line of separation between the land-cloud and the sea-cloud." Whatever caused that line may have remained uncertain, but the line itself was not. The line was, in fact, a stable and even celebrated feature of the landscape. When the great Prussian explorer Alexander von Humboldt had stopped at the island in 1799 at the beginning of what would become his epic five-year journey to South America, he too had noticed the curious phenomenon by which the clouds parted to reveal the top of the peak.[2] Some thirty years later, Charles Darwin had witnessed the same meteorological unveiling when he visited the island in January 1832 at the beginning of his own voyage to South America. He mentioned it in his *Naturalist's Voyage*, remarking how "we saw the sun rise behind the rugged outline of the Grand

The Cloud Horizon Westward from Guajara, shewing the summit of Palma above and the base of Gomera below the Cloud

FIG. 3.1. The peak of La Palma seen above the cloud line in a drawing by Charles Piazzi Smyth in 1856. Alexander von Humboldt and Charles Darwin had observed and written about the same meteorological feature. Credit: Royal Observatory Edinburgh.

Canary Island, and suddenly illuminate the Peak of Teneriffe, whilst the lower parts were veiled in fleecy clouds."[3]

In his own narrative describing his journey to test the possibilities of mountaintop astronomy, Piazzi Smyth mentioned the existence of a scientific explanation for the delineated cloud lines before remarking that at the moment he caught sight of the peak through the clouds, "the effect on the feelings was such, that there could have been few persons with whom the leading idea would have been the physical explanation." Clouds and their motions evoked feelings more readily than thoughts, suggested Piazzi Smyth. Or did they? Piazzi Smyth was coy about whether he was one of the people for whom physical explanations did in fact dominate. He claimed that awe and wonder preceded scientific understanding, but in the telling, Piazzi Smyth is scientist first, wonderer second.

That the clouds, as the most visible and visibly changeable aspect of the weather, might produce strong emotions had the self-evident truthfulness of cliché for Piazzi Smyth. In the early decades of the century, the English painter John Constable had offered a newly prominent role for the sky—formerly mere backdrop—as "the keynote, the standard of scale, and the key organ of sentiment" in landscape painting.[4] By sky, he really meant clouds. In his finished paintings and in his remarkable series of cloud studies, he single-handedly transformed clouds into the primary pictorial mechanism for delivering emotional impact. This did not mean banishing the techniques of science from art. On the contrary, scientific techniques for Constable worked in the service of emotional veracity. He believed that a large portion of the "artistic" quality of a work he painted lay in the authenticity of the emotions it provoked. Did it make the viewer feel as if she were standing in a field, watching the scene unfold before her? If scientific tools and practices could increase the emotional impact of a painting, this was all to the good.

Constable, the consummate artist, had learned to see the clouds partly through the eyes of Luke Howard. When in 1803, Howard had offered a newly standardized nomenclature for clouds, he provided a new set of tools for capturing the felt reality of clouds, in words and pictures. For Constable and the painters he influenced, scientific understanding of the clouds could be utilized to generate a convincing subjective experience. For scientists, the trick was to devise methods for describing clouds that captured not the feelings produced by clouds but their changing nature. As important as Howard's imposition of order on what had previously seemed impossibly disordered was his insight into the ways in which clouds were transformed into other clouds. The study of changing types, rather than fixed forms, was written into his project from the start. The role of emotion in the service of Howard's science was less certain. What was clear, instead, was how difficult it was to separate the two. Clouds were interesting, useful, and important precisely because of the ways they elided—glided across—the boundary between objectivity and subjectivity, between science and art, between fact and feeling.

In the same way that Tyndall and Forbes's dispute over glacier motion hinged not on who had the better explanation but on what counted as an explanation, so the men who wished to study clouds scientifically were at pains, in 1856, to define what exactly that might mean. In 1804, Howard had provided one explanation—to *know* a cloud meant to identify it and to name it. This natural historical approach treated clouds as specimens of nature that could be observed and collected in much the same way as butterflies. And just as biologists could tell a great deal about butterflies by their taxonomic descriptions, so would it be possible to learn much about the geography of clouds by this technique. Though Howard emphasized how important it was to attend to the modifications of clouds, he made no suggestions about the physics of cloud transformation, nor about the role of clouds in the generation of storms.

By 1856, clouds were increasingly subject to a new sort of scrutiny and to new ways of being known. A new government office for weather, called the Meteorological Department, was established in 1854 with the intention of increasing knowledge about the weather for both practical and scientific benefit. The doubled mission of the office was evident in the choice of an Admiralty captain as its first head. Admiral Robert FitzRoy, who had captained the *Beagle*, the ship upon which Darwin had served as naturalist (and from which he had observed the clouds at Tenerife), was a practical navy man whose interest in clouds was to protect the sailors who served under him, and, by extension, any Briton who might come to harm during a powerful and unexpected storm. While government bureaucrats and scientists both agreed that reliable predictions of coming weather were far off, FitzRoy took a pragmatic approach to the matter. He thought that it was more important to use knowledge of the weather to save lives than to wait until a "mature" science could be established on the basis of statistics. This would lead to both pathbreaking and highly controversial action on his part.

Piazzi Smyth was thirty-seven when he arrived at Tenerife.[5] It was an expedition toward which his whole life had led him. He had been born under a Neapolitan sky and christened with a name like a prophecy: Charles Piazzi Smyth. Sandwiched between solid Scottish nomenclature nestled the surname of Giuseppe Piazzi, a great Italian astronomer and friend of his father. The exoticism and ambition, of both his Italian namesake and his own father, himself an accomplished naval officer, both took root in the child. By the age of sixteen, he was well on his way, literally and metaphorically, as he departed the Bedfordshire school where he'd been studying and sailed along the west coast of Africa until he reached the farthest point of the continent. He landed at the Cape of Good Hope in 1835 where, by prior arrangement, he was to spend the next ten years as an apprentice to the Royal Astronomer at the Cape, Thomas Maclear.

He learned how to locate and precisely map the stars—vastly more of them visible in the dry air of the Cape than in Britain. He worked hard in helping measure the length of an arc of the meridian. He spent parts of five winters surveying the land in the pursuit of geodetic precision, enduring cold mist and icy winds in the mountains of the western Cape. He traced the faint, dusty glow of the zodiacal light, scattered by dust held in the plane of the solar system, a delicate, elusive phenomenon that hovered on the edge of visibility.

Seeing required training, and so did recording what he saw. He sketched from an early age and developed a fluent, veridical style. He drew the view from his room at the school in Bedford, the people on board the ship that carried him to the Cape, and the buildings that he encountered once there. He sketched Halley's Comet as it passed in 1835–1836. He was the first person to make photographs in Africa, experimenting with making rudimentary images of plants even before he had learned which chemicals to use. As early as 1843 he was able to take images that survive today—people and buildings in southern Africa, including one of the Cape Observatory (possibly the oldest photograph of any observatory).

He returned to Britain to become Astronomer Royal for Scotland at the age of twenty-seven. It was a precocious appointment, but he soon realized that the grandest thing about the post was the title. He had scant funds, and the observatory was chronically understaffed. The skies of Edinburgh were smeared with hazy coal smoke, layered over with low cloud, gray as the stones on the houses of New Town. It was hard to see anything like what he had seen in the Cape. Nevertheless, he set himself to fulfilling the demands of his office.

His moment of inspiration arrived at the same time as a wife. He decided to embark upon a journey to Tenerife, to see if he could bring precision instruments to the top of the mountain and establish an observatory there. No longer precocious, by this time Piazzi Smyth was thirty-six, and his bride, Jessie, a surprising forty. They married on Christmas 1855, and by the following June they were sailing to Tenerife. In the hold of an expensive yacht that transported them were the following instruments: an actinometer, magnetometer, thermometers, electrometers, spectrum apparatus, and polarimeter, loaned by none other than the Astronomer Royal, George Airy. Barometers and more thermometers were loaned by Admiral FitzRoy, head of the Met Office. The hydrographer loaned him four chronometers. And Robert Stephenson trumped them all by lending him an entire yacht, the RMS *Titania*, with a crew of sixteen men for the return journey.

It was a classic voyage of imperial reckoning, made possible by the well-engineered tools of the industrial revolution, the instruments he'd been loaned by the greatest scientists of the day, the expensive sailing ship, and the well-trained crew he had the run of. What Piazzi Smyth was doing was attempting to verify an old hypothesis with an enviable pedigree. Isaac Newton had proposed in his *Opticks* of 1704 that astronomical observations would be greatly improved by removing the "injurious portion of the atmosphere." Since then, many had concurred but no one had attempted to prove the point. Tenerife was closer to London than the Cape, and so more convenient, but it also presented potentially insurmountable obstacles to scientific observation. It was possible that the instruments would prove impossible to transport to the peak, fail to operate once there, or that perpetual cloud would surround the summit. If, on the other hand, those obstacles could be overcome, then more science and more scientific vision could be had.

The mountain could be a machine for making facts out of theories, as Piazzi Smyth put it. And so, by implication, could the astronomer himself. But doing so required balancing between worlds in a manner that brought to mind a man teetering atop a peak. A Scot who had been born in Naples, trained in southern Africa, and pro-

FIG. 3.2. The crew of the *Titania* en route to Tenerife, photographed by Charles Piazzi Smyth in 1856. Credit: Royal Observatory Edinburgh.

fessionally employed in Edinburgh (a proud capital that was simultaneously an outpost of London), Piazzi Smyth was a creature of the periphery. As such, Piazzi Smyth was uniquely qualified to make the attempt.

His success depended on maintaining the standards of metropolitan astronomical science on a mountaintop several thousand miles from Britain. His own excitement, and the appetite for exploration it fed, had brought him first to the Cape and now to Tenerife. It was a necessary precondition for an explorer-scientist in the middle decades of the nineteenth century. But the spirit of exploration sometimes sat uneasily with the sort of self-restraint insisted upon by the men who stayed in London. For while the requests of the London scientists who had loaned him the instruments (and given him the money) were many, and varied, they were also quite firm—even rigid—about the extent of the domain to which they felt Piazzi Smyth had earned access. Even on an expedition designed to go further, astronomically speaking, than had ever been gone before, it was possible to go too far. That went for the types of observations Piazzi Smyth could make—geology and biology were not welcomed—and it also went for the forms of expression Piazzi Smyth used to describe his discoveries. Piazzi Smyth knew this, and this knowledge helps explain the defensive note that crept into his writing about the moment of first contact between him and the mountain. He knew that he had to suppress his own emotional response in favor of the instrumental readings he'd been empowered to make. If he worked hard and got lucky, he might manage to render the mountain a suitable outpost for British astronomy, a new sort of colony: a temporary, provisional, but potentially bounteous source of new knowledge. This goes some way toward explaining Piazzi Smyth's curious locution and defensive recounting of his dramatic arrival at Tenerife. He was trying to abide simultaneously by two norms for watching the skies—the one based on emotion, the other on what he called "physical explanation." What is interesting about Piazzi Smyth is not only that he felt himself to be caught between—or spread across—these two ways of looking at clouds, but that he shared the experience of a doubled response with his readers.

The need to eliminate the personal from scientific observation took on a special urgency in the case of mid-nineteenth-century astronomy, when it became apparent that differences in the reaction times of observers could become a significant source of error when it came to making extremely precise observations of celestial movements. There was a phrase for this problem, the "personal equation," which suggested the desirability of reducing human differences to a numerical factor, a handicap that could then be subtracted from the results, giving a true number. Astronomers became professionally paranoid, on guard against any and every source of error. "Vigilance can never sleep; patience can never tire," wrote one popular writer at the end of the century. "Variable as well as constant sources of error must be anxiously heeded; one infinitesimal inaccuracy must be weighed against another; all the forces and vicissitudes of nature—frosts, dews, winds, the interchanges of heat, the disturbing effects of gravity, the shiverings of the air, the tremors of the earth, the weight and vital warmth of the observer's own body, nay, the rate at which his brain receives and transmits its impressions, must all enter his calculations and be sifted out from his results."[6]

Even taking such extreme precautions, it was impossible to eliminate personal differences between observers—specifically, the reaction times that varied between observers trying to determine with extreme accuracy the time at which a star crossed a certain location in the sky. The more precisely astronomers were able to map the stars, the more the personal equation mattered, since small differences in the reaction times of observers made a big difference when tiny units of time were being measured. One way around this problem was to establish hierarchies of observers, each of whom was himself observed by the managers of observatories, men such as George Airy, head of the Royal Observatory at Greenwich.[7] Piazzi Smyth was not alone on the mountaintop, but from the perspective of astronomers like Airy, he might as well have been. There was no one to watch Piazzi Smyth, no one against whose observations his own could be checked, no one observing him making his observations.

When Piazzi Smyth mentioned that for most people, looking up at a sublime bit of meteorology—the shifting of clouds to reveal a monumental peak—was an emotional experience, he was, in a somewhat oblique way, making reference to a perennial astronomical bugbear. Though astronomers used the term *personal equation* in hopes of eliminating differences between observers, Piazzi Smyth was here drawing attention to the ineffability of personal observation, to the way it could not be reduced to numbers. By rendering the subjectivity of the observer into a commonplace to be acknowledged rather than a troublesome anomaly to be eliminated, Piazzi Smyth was suggesting the possibility that scientists could be simultaneously objective and subjective, impersonal and personal.

If the problem of precision was significant in astronomy, it was partly because the astronomy that Piazzi Smyth and most of his contemporaries were doing was, above all, a cartographic exercise. The vast energies and expenditure poured into astronomy in the first decades of the nineteenth century by the French and the British were a form of scientific colonization. Nearly a century after Newton had shown it was possible to predict the motions of heavenly bodies according to a set of physical laws, astronomers were mostly still preoccupied with working out what this meant in practice. Mapping the position of the stars, the sun, the moon, and the planets—called positional astronomy—was a continuation of the research program that Newton had first set out in 1687 with the first edition of his *Principia*. For this, long hours of observation, with the most precise instruments, used by the best-trained observers overseen by the most demanding supervisors, were necessary to produce a map sufficiently refined to demonstrate both the theoretical potential of the Newtonian system and, just as important, to wring from it the practical benefits of improvements in navigation and surveying. Knowing the heavens made it possible, in a direct and practical way, for nations to know the earth, and in knowing the earth, to control an ever-greater part of it.[8] It also made it possible to predict with extraordinary accuracy the movements of celestial objects, an accomplishment which brought considerable prestige to the discipline of astronomy and served as a beacon for the ambitions of many other physical sciences.

As powerful a tool as positional astronomy could be, astronomers always hoped for more. Piazzi Smyth had come of age in the 1830s, when astronomers felt increasingly emboldened to hope that it might be possible to say not only where but also *what* the stars were. Once astronomers saw the prospect of moving beyond celestial mechanics—once seen as the ultimate "perfect" system—an exciting but newly confounding world opened up before them. Newton's cosmos had been sterile—a clean clockwork universe in which God made intermittent appearances to keep the planets on their eternal orbits but little else occurred. The new cosmos was filled with energy that bombarded the planet, bathing it in a relentlessly dynamic flux of light and magnetism. The smooth and singular orbits predicted by Newton's mathematics were replaced by complex and messy traces of thousands of barometers, thermometers, magnetometers, and a host of other instruments that sought to catch the cosmic fluxes of the universe.

There was no more influential proponent of the idea that nature could be made to yield her secret laws than Alexander von Humboldt, the Prussian explorer and naturalist. When he stopped at Tenerife, he laid his anchor in mist "so thick, that we could scarcely distinguish objects at a few cables' distance." Like Piazzi Smyth, he feared that the mountain would remain out of sight, but "at the moment we began to salute the place, the fog was instantly dispelled. The Peak of Teyde appeared in a break above the clouds, and the first rays of the sun, which had not yet risen on us, illuminated the summit of the volcano."[9] Despite the mist, Humboldt noted the effects of the transparency of the atmosphere, "one of the chief causes of the beauty of the landscape under the torrid zone." Not only did it heighten colors, harmonizing and contrasting them, but it changed the very "moral sensibilities" of the inhabitants of southern realms, leaving them with a "lucid clearness in the conceptions, a serenity of mind, correspond[ing] with the transparency of the atmosphere."[10] Clear skies, then, could lead to clear minds.

Humboldt never stopped thinking about both how natural environments affected human beings and how humans could understand the physical nature of those environments. Decades later, as he sat down to write a book that was the culmination of a lifetime of travel and contemplation, he returned to the question of how different landscapes affect people differently, or the "different degrees of enjoyment presented to us in the contemplation of nature." Thinking back on the many places he'd been, a few leapt out. In addition to the "deep valleys of the Cordilleras," where tall palms

FIG. 3.3. Alexander von Humboldt, painted by Julius Schrader in 1859, with Mount Chimborazo and Mount Cotopaxi in the background.

had created a forest above the forest, he returned once again to Tenerife, remembering

> when a horizontal layer of clouds, dazzling in whiteness, has sepa-
> rated the cone of cinders from the plain below, and suddenly the
> ascending current pierces the cloudy veil, so that the eye of the
> traveler may range from the brink of the crater, along the vine-clad
> slopes of Orotava, to the orange gardens and banana groves that
> skirt the shore.

What was it that gave scenes like these the ability to move a man's heart, to spark the "creative powers of [a man's] imagination"? Part of their potency lay in their changing aspects, in the way moving clouds or water dramatized the flux of forces that was ever present but not always so visible. Humboldt called this the "peculiar physiognomy and conformation of the land, the features of the landscape, the ever varying outline of the clouds, and their blending with the horizon of the sea." The result of all this change was that he, like Tyndall, had the eerie sense that nature was imbued with emotion—his own emotion. "Impressions change with the varying movements of the mind," wrote Humboldt, "and we are led by a happy illusion to believe that we receive from the external world that with which we have ourselves invested in it."[11]

This happy illusion was in large part due to the unity of Nature. "The powerful effect exercised by Nature springs, as it were," wrote Humboldt, "from the connection and unity of the impressions and emotions produced." Unity was the feature that caught the attention. To achieve true understanding, it was necessary to go deeper. As humankind developed intellectually, it became possible to move beyond the primordial feelings of unity and arrive at an even more powerful method for apprehending the world.

By degrees, as man, after having passed through the different gradations of intellectual development, arrives at the free enjoyment of the regulating power of reflection, and learns by gradual progress, as it were, *to separate the world of ideas from that of sensations*, he no longer rests satisfied merely with a vague presentiment of the harmonious unity of natural forces; thought begins to fulfill its noble mission; and observation, aided by reason, endeavors to trace phenomena to the causes from which they spring. [emphasis added][12]

By separating thought and emotion, Humboldt thought it would eventually be possible to disentangle the many threads of differing phenomena—magnetic, astronomical, and meteorological—and assign them to their respective origins, or, in other words, "to trace phenomena to the causes from which they spring." Only by regulating emotion could mankind surpass the powerful first impression of unity. Humboldt's vision was both gradual and bold. It would take time, but in the end a much deeper understanding of the multiple forces of nature at work could be achieved. The way to turn "mere" natural history into a physics of the earth (*Physik der Erde*) was to "track the great and constant laws of nature manifested in the rapid flux of phenomena, and to trace the reciprocal interaction, the struggle, as it were, of the divided physical forces."[13] Taken together, these readings would reveal the true and truly singular face of the earth.[14]

Just as it became increasingly possible and attractive to determine the effect of physical forces on earth, so it became almost irresistible to look for the invisible but powerful threads that linked the earth and the heavens. Humboldt did not distinguish between the forces present on earth and those that prevailed throughout the universe. His approach encompassed nothing less than the entire cosmos. The "harmonious unity of nature" necessarily knit together the heavens and the earth. In the same way, it constituted a sort of undulating tapestry of physical forces that could be discerned in the isolines of temperature and pressure, and in the corresponding bio-geographical continuities that Humboldt so painstakingly reconstructed.

Humboldt's vision was shared by John Herschel, the son of the great astronomer William Herschel, discoverer of Uranus, and himself an accomplished scientist. John Herschel helped organize the so-called Magnetic Crusade of the 1830s, an ambitious attempt to simultaneously map changes in the earth's magnetic field at different places around the globe.[15] The results of this multi-year expedition were stunning, revealing that Earth's magnetic field changed in response to that of the eleven-year cycle of sunspots. No better example of Humboldt's belief that beneath flux lay order could have been hoped for. Here was a powerful justification both for the gathering of multiple sorts of data—on sunspots, solar spectra, gravity, radiation, and much besides—and for the effort needed to disaggregate those phenomena from each other. Knowledge, in this sense, would be as much subtractive as additive. To understand both the emotional power of a place like Tenerife, and the physical phenomena made manifest, required tools for working backwards from the powerful impression of unity to the underlying causes which together operated to create the traces recorded by the many instruments Piazzi Smyth took with him to the island.

For Piazzi Smyth and his peers, separating the effects of the terrestrial atmosphere from that of the solar atmosphere was a prerequisite for understanding the true nature of either. In this sense, it was not possible to do solar physics without doing terrestrial physics or vice versa. This new physical way of doing astronomy firmly knit together the earth and the cosmos, to create, as one writer put it, "a science by which the nature of the stars can be studied upon the earth, and the nature of the earth can be made better known by study of the stars—a science, in a word, which is, or aims at being, one and universal, even as Nature—the visible reflection of the invisible highest Unity—is one and universal."[16]

Earth's atmosphere played a peculiar role in this search for unity. In its shifting movements, the atmosphere and the clouds which floated in it represented the difficulty of seeing to the essence of things, as well as the need to always be vigilant about the objective of observation. Clouds were sometimes obstructions to the visions that lay beyond—of stars, of peaks. But they were also aspects of the natural world and therefore worthy of study in themselves. They represented a particular variety of the fullness of nature, the way in which the veils that she drew across herself were themselves part of nature. The objects that obscure vision are themselves worth looking at. In this way, the atmosphere was a doubled object, both an impediment to science and an object of it.

How did Piazzi Smyth, in practice, set out to disentangle the myriad forces that together made up the unity of Nature? The first, elusive glimpse of the peak, revealed through the clouds, pointed the way. His objective was to get as high up the mountain as was possible and locate a place on which an observatory could be built. A crew of twenty porters and twenty mules helped him get there, once the cumbersome boxes he'd brought on the ship had been broken down into parcels more suitable for carrying up a jagged volcano than stowing on a ship. By one o'clock on the first day of climbing, the expedition reached over 7,000 feet. "Light and heat revel everywhere," marveled Piazzi Smyth. "There is no need of volcanic assistance."[17] They reached the top of the mountain's intermediate peak, Guajara, as day fell. The porters hastily stashed their loads and sped back down the mountain to spend the night at a more reasonable altitude. Piazzi Smyth and the rest of the small party of foreigners stayed at the summit. Piazzi Smyth exulted that he was, "within twenty-four days of leaving England, bivouacking at a height of nearly 9000 feet, on a mountain only 28 degrees from the Equator."[18]

On the way up the mountain, they hiked through the discrete layer of clouds he'd seen from below on the ship. The layer floated like an atmospheric sea, above which the peaks of Tenerife and nearby La Palma jutted, secondary islands emerging not from the liquid water of the ocean but from the condensed water droplets that formed the cloudscape. Those clouds were persistent and uniform, and they stretched as far as the eye could see. They were a literal and a metaphorical boundary. Piazzi Smyth called it "that great plain of vapour floating in mid-air at a height of 4000 feet." It was a "separator of many things. Beneath were a moist atmosphere, fruits, and gardens, and the abodes of men; above, an air inconceivably dry, in which the bare bones of the great mountain lay oxidizing in all variety of brilliant colours, in the light of the sun by day, and stars unnumerable at night."[19] Above, too, was the realm in which lay the astronomical justification for his trip, the vindication of Newton and a sharpness of celestial vision otherwise inconceivable.

What clouds were and to which landscapes they belonged was an open question. Howard had altered the form and content of meteorology when he had suggested that clouds were neither endlessly variable nor unclassifiable. But many questions remained, not least whether clouds were, like living species, indigenous to certain places or whether they were more general, universal features to be found throughout the globe. When Piazzi Smyth went up the

mountain, the clouds at Tenerife were noteworthy for their difference from English clouds, but they were also, it was to be hoped,
potentially able to be reduced to universal laws just as the changing
magnetic readings taken during the Magnetic Crusade had been.

The intensity of the light changed the nature of time on the mountain. Piazzi Smyth could see so much more in a single day or night
than he could ever see in lower realms that he was able to achieve an
astonishing amount of observing. "The day wears apace," explained
Piazzi Smyth, "and most luxuriantly in so pellucid an atmosphere,
lit up by the rays of a vertical sun, undiminished by any aerial impu-
rity. Each moment on a day of this sort is worth hours on any other;
we look at everything far and near, see it as it were face to face,
and gain a higher idea of the glorious creation in which we live."
Color acquired extraordinary depth: "glorious cadmium," "the rich-
est tint of red-orange," "lemon-yellow," "powerful rose-pink" and,
finally, the "deep blue sky above."[20]

Much work was needed to transform the light into a useable
scientific tool. Piazzi Smyth spent about a month camped at 9,000
feet before frustration with persistent dust sent him higher, to the
appropriately named Alta Vista, at 10,700 feet. He set about doing
what he had avoided doing at Guajara, transporting the "great Pat-
tinson equatorial telescope," which required "straining every nerve
to accomplish the main feature of the expedition—viz. to place the
largest telescope on the highest available part of the mountain."[21]
Around the inner "telescope square," a group of local men and some
crew members from Stephenson's ship labored to create a group of
five rooms (with roofs, Piazzi Smyth noted proudly) and a veranda
which offered some protection from the elements.

Down to its bones, the place was hybrid. The walls were built
of rocks from the summit, to which were added felt wall coverings
and supporting timbers made by young poles of fir brought from
Tenerife and glass plates, shutters, and door hinges brought all the
way from Edinburgh. Plain nails were plentiful on the island, but
"good screw nails," noted Piazzi Smyth, "seem to be bound up with
the march of Anglo-Saxon civilization."[22] It was a joke that showed
how much the success of the astronomical experiment depended

FIG. 3.4. Jessie Piazzi Smyth with telescope and sun hat at the peak of Alta Vista. Credit: Royal Observatory Edinburgh.

on reproducing conditions back in Britain, down to the screws used to fix the instruments together.

It wasn't easy. Most of the more than 500 pages of Piazzi Smyth's book on the subject were spent describing just how hard it was. His tone was not querulous but full of amazement. "Some part or other of our photographical apparatus, for picturing the sun's image," he explained, "would every now and then begin to smoke and burn." The eyepieces to the telescopes became dangerously heated, so they periodically had to stop in order to keep from burning themselves.[23]

It was worth it, though, and he knew it from the start. The difficulties were not avoidable. In fact, they were necessary. His first effortless, instantaneous gaze depended not only on the absence of vapor from the atmosphere but on an unbroken chain of labor and supervision (of men and materials) stretching from the summit back to Edinburgh and London. All of this hard work—the packing and portaging, the building and the training—become as transparent as the atmosphere the moment Piazzi Smyth put his eye to the telescope's eyepiece. That is the magic trick behind this kind of scientific work—behind, in a sense, all scientific work: a great amount of work is applied to making a small bit of nature visible in a way it has never been visible before.

When Piazzi Smyth aimed his telescope at the sky and set his eye against the eyepiece, a singular line of vision stretched from there to the farthest star. He could see farther, much farther, than he (or anyone) had ever seen before. This is worth repeating: From his vantage on the mountaintop, armed with a powerful telescope, with the clear air stretching above him to the emptiness of space, Charles Piazzi Smyth could see farther into the heavens than anyone had ever seen before. In the first night of observing on the summit, he transcended the viewing records of a lifetime. Pairs of stars, normally blurred and indistinct, leapt out at him. Even the faintest stars, of sixteenth magnitude, were easily visible. He quickly ran out of astronomical tests with which to gauge the extent of the vision he'd acquired.[24]

Having proven the practicality and desirability of mountaintop astronomy, Piazzi Smyth set about taking observations that would contribute to the exciting developments in physical astronomy, in showing *what* the stars and the planets were, rather than just where. The instruments he'd so laboriously transported offered the potential to do what Humboldt had urged and begin to tease apart the physical phenomena that together made up our impressions of both heaven and earth. What caused the spots on the sun to change, and by what cycle did they do so? What were the red prominences that shot out from the sun, visible during eclipses but presumably present all the time? What was the nature of the double stars, and how did their rotation change over time? What patterns controlled the tides, the earth's weather, its magnetic field?

There were so many questions. It was impossible to answer them all. But the fact that they were being asked at all indicated how much had changed in the way people thought about the heavens and the earth. New and improved instruments made it possible to "see" invisible physical phenomena for the first time. Increasingly powerful telescopes were able to gather more light and resolve finer detail. Photography was put to astronomical uses almost as soon as it was invented, when Louis Daguerre pointed a camera at the moon in

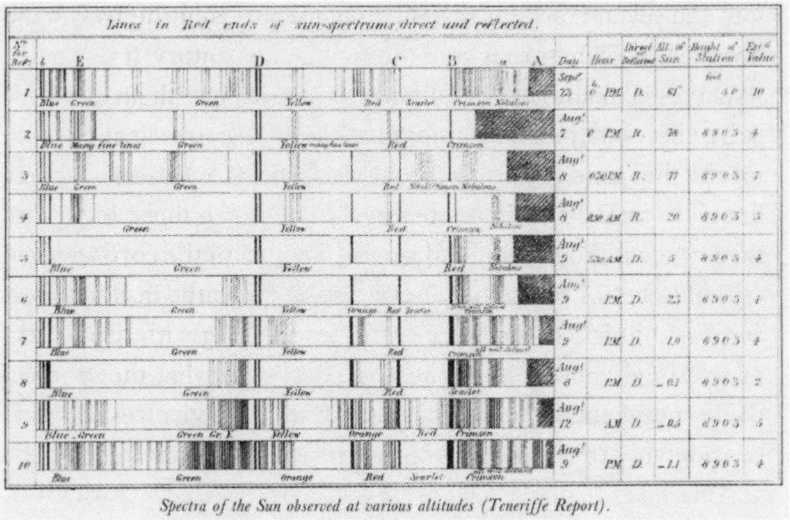

Lines in Red ends of sun-spectrums direct and reflected.

Spectra of the Sun observed at various altitudes (Teneriffe Report).

FIG. 3.5. Spectra of the sun observed at various altitudes and times of day from Charles Piazzi Smyth's report on the Tenerife expedition. The bottom reading was taken when the sun was setting.

1839, a feat repeated with more success the next year by John William Draper, who devised a way to track the moon during the long exposure. Images of the sun followed in the 1840s, and the first star, Vega, submitted to photography in 1850. Most influential of all was the spectroscope, a device that made light an experimental tool capable of diagnosing the contents of distant atmospheres. It provided yet more evidence for the unity of nature, proving that the same elements were present on earth and in the heavens.

The distinctive array of colors produced by the diffraction of light had been observed for centuries. Leonardo da Vinci had noted "rainbow colors" around the edges of air bubbles in a glass of water. Isaac Newton had introduced himself to scientific society when he'd demonstrated that the oblong of light produced by a prism of sufficiently clear glass was made up of colors that could not be further modified. He coined the term *spectrum*—playing on its double meaning as both a ghostly image and something which is seen—to describe the rainbow of colors which were refracted to different degrees by the prism. It was Newton who divided the spectrum into the seven colors, a system which dominated the way people saw the spectrum throughout the eighteenth century. It was only in 1802 that William Hyde Wollaston, a physician with an interest in light, observed the spectrum through a very narrow slit and noticed for the first time that atop the spray of colors lay a series of black lines. He made an initial attempt to map these lines, identifying the five most prominent, and labeled them with the corresponding capital letters *A* through *E*. Acting independently, in 1824 Joseph von Fraunhofer, a German glassmaker working with high-quality prisms (and consequently more interested in what the spectrum could reveal about the purity of the glass than vice versa), added considerably to the map, assigning unique numbers to more than 500 lines.

Seeing the spectrum was no simple matter. No one knew how many lines there might be. The closer anyone looked, the more seemed to appear. Still less certain was what caused them. This made it that much harder to know when to trust one's eyes. A further complication was the difficulty of translating what was seen into something that could be represented graphically. Piazzi Smyth, trained from an early age in just these skills, was exquisitely attuned to just how much skill was required to accurately represent astronomical phenomena with techniques such as John Herschel's use of "a fine camel's hair brush" and successive washes of varnish to draw stars. A faithful imitation of such tricky phenomena to record as the Aurora Borealis, a cloud of nebular gas, or the tail of a comet could only be obtained by "correctness of eye, facility of hand, and a due appreciation of the subject."[25] The spectrum, with its spray of lines of varying strength, which seemed to come and go, was especially hard to capture.

Charles Babbage, impresario of science and decrier of British decline in relation to the French and the Germans, not only recognized that vision was a skill that had to be acquired, but saw that it had national repercussions. In his *Reflexions on the Decline of Science in England*, he recounted how Herschel had warned him how difficult it was to see spectral lines of the sun. He could sit in front of a spectroscope through which the solar lines were visible, said Herschel, but Babbage would not be able to see them until he had been told "how to see them," at which point, and only at which point, he would then be able to see them. After having seen them, Herschel told Babbage, you will wonder how you could have missed them and will never be able to look at a spectrum again without seeing them. And so it was.[26] Without good systems for training observers, Babbage concluded, British astronomy could not compete on the international stage.

The drive initially was to map more and more lines. Soon, it became clear that the number of lines visible depended not only on the size of the telescope or the clarity of the prism, but on the time of day and direction in which the telescope was pointed. In 1833, the physicist David Brewster announced the results of a multi-year project. Not only had he observed the spectrum at a resolution some four times greater than what Fraunhofer had achieved, but he had observed the spectrum at different times of the year, under different meteorological conditions, and with the sun at varying angles in the sky. With his observations, Brewster was able to move toward the Humboldtian dream of disaggregating phenomena in order to understand them better. In 1856, when Piazzi Smyth embarked for Tenerife, it still remained very unclear what the cause of these lines were and where they originated.

And so, in addition to his nighttime observations of the stars, Piazzi Smyth used the spectroscopes he'd brought to study the characteristic lines that appeared when sunlight was observed passing first through a slit and then through a prism. The metropolitan scientists, tethered to their urban observatories, wanted to know if lines appeared or disappeared when Piazzi Smyth looked through the spectrum on the mountaintop. Did they look different at sunset or sunrise?

Here the mountain came into its own as a tool for separating the earth's atmosphere from the sun's. Atop the peak, observing sunlight with his spectroscopic apparatus, itself a hybrid of telescope, prism, and a slit that spread the spectrum and thereby revealed the lines, Piazzi Smyth was uniquely positioned to help solve the question. Pointing the spectroscope at the sun at midday, he was closer to the sun's atmosphere than any other similarly equipped observer on the planet. So too, as he tracked the sun on its ascent and descent, at dawn and sunset, when it hovered just at the horizon, he had access to a thicker portion of the atmosphere than anyone else on the planet.

As with the distant stars, so with the sun, what Piazzi Smyth saw from his privileged position seemed effortlessly dispositive, unmistakably clear. As he watched the setting sun through the apparatus, he saw the number of lines grow visibly before his eyes. This was evidence that at least some of the lines were earthly in origin, a visible marker of some invisible substance which increased as the section of the earth's atmosphere he was looking through thickened. This meant that the spectrum revealed by any spectroscopic apparatus pointed at the sky was always a hybrid, a representation of the atmospheric contents of both the sun and the earth. This complicated the quest to identify the contents of something so distant as the sun with little more than a special piece of glass. But Piazzi Smyth's observations on Tenerife also showed that the spectroscope, used in the right location and with the right techniques, could be a tool for revealing the differences between those contents and for plumbing the atmospheric reaches of the earth itself.

What caused the lines that grew before his eyes, Piazzi Smyth did not hazard to guess. Nor did he wonder, at that moment, about the precise details of their variation, whether the lines ebbed and grew solely due to the amount of atmosphere through which he was observing or whether the internal changes of the earth's atmosphere itself affected the pattern he saw. Those thoughts would come later. At the time, the observations formed just one part of the dozens he was making, using every waking moment on the mountain.

We know how Piazzi Smyth felt as he craned his neck to look up at the summit of Tenerife because he wrote a book about his experience. His great subject was the act of observation itself, and hidden in every record of the outside world that he made over a lifetime of watching was a record of Piazzi Smyth, the observer. Convinced of the advantages of photography for scientific observation, Piazzi Smyth spent "all spare moments" on the mountain with a camera he had gotten at the last minute, making images of the surrounding landscape, the unusual flora, and the work of scientific observation itself. The description I have recounted takes up about five pages in *Teneriffe, An Astronomer's Experiment: Or, Specialities of a Residence Above the Clouds*, which Piazzi Smyth published in 1858. That book provides a lengthy description of the whole voyage in language that is colorful without being florid. It also contains a set of twenty stereo-photographs taken by Piazzi Smyth on the expedition, the first time such a set of photographs had been included in a printed volume.

In his preface to the book, Piazzi Smyth explained the reason he had taken such care and effort to make and print his doubled photographs: They possessed what he called a "necessary faithfulness." While single photographs may contain smudges or artifacts, the stereo-photograph can serve as its own correction. A comparison of the two images will reveal what is real, and what is merely accidental. A further degree of veracity is achieved when the images are combined stereoscopically to produce the impression of distance or solidity, normally the purview only of great painters. The doubled images themselves do a double duty, producing scientific accuracy and the kind of aesthetic "effects" normally produced by artists.

Piazzi Smyth was always doing science and watching himself (and others) do science. He recorded natural phenomena and the process of recording natural phenomena with the same level of interest. This meant that he sketched the boat (itself a scientific instrument of exploration) upon which he traveled and even took photographs of the crew in the difficult setting. It meant that he kept a detailed journal of the expedition (a long-standing dual form of registration— both of the natural world and of the perceptions of the observer of the natural world). It meant that he carefully observed the sailors from his ship who had transformed themselves (with his help) into disciplined observers themselves. He included a photograph of the second mate of the ship in the act of taking a temperature measurement. In one hand is the chronometer he used to time the observation; in the other is the notebook in which he recorded it. The image, like the others in the book, appears in stereo—a further dou-

FIG. 3.6. Stereo-photograph image of ship's mate making observations, taken by Charles Piazzi Smyth. Credit: Royal Observatory Edinburgh.

bling of an already doubled act of observation: Piazzi Smyth watching the second mate watching the temperature. Similarly, the act of scientific observation was itself carefully watched here, not only by Piazzi Smyth but also by the readers of his popular and his official accounts. Science, an act of observation, required observation itself to be regulated.

The two photographs Piazzi Smyth chose to include in his official report were both images of the summit of the mountain. The first was a stereoscopic portrait of something no one had seen, including Piazzi Smyth himself. It was a double photograph of a model of the summit made by the engineer and talented amateur astronomer James Nasmyth. Piazzi Smyth photographed the model, based on data collected by the expedition, from above, and printed it in stereo view, providing a pure image of what the crater would look like to those with perfect vision and a perfect, God's-eye vantage point. The second photograph was a single image (made from an enlarged camera-copy) of the Alta Vista Observatory, also taken from above. This view was a "real" view in the sense that Piazzi Smyth had climbed up a nearby peak from which he was able to look down on

REPORT

<small>ON THE</small>

TENERIFFE
ASTRONOMICAL EXPERIMENT

OF 1856,

ADDRESSED TO THE LORDS COMMISSIONERS OF THE ADMIRALTY,

<small>BY</small>

Charles
Prof. C. PIAZZI SMYTH, F.R.SS. L. & E., F.R.A.S., AND

H. M. ASTRONOMER FOR SCOTLAND.

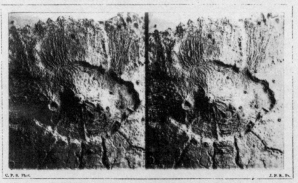

C. P. S. Phot.   J. P. S. Pr.

STEREOSCOPIC MAP OF THE PEAK AND GREAT CRATER OF TENERIFFE.
From a Model by J. Nasmyth. Esq,. C.E., founded on data procured by the Expedition.

*/ LONDON AND EDINBURGH:

PRINTED BY RICHARD TAYLOR AND WILLIAM FRANCIS, RED LION COURT, FLEET STREET, LONDON,
AND NEILL AND COMPANY, HIGH STREET, EDINBURGH.

1858.

FIG. 3.7. Title page to Charles Piazzi Smyth's official report of his expedition to Tenerife, with a stereo-photograph by him of a model of the peak by James Nasmyth.

the observatory. The image shows the arm of the great telescope emerging from the "telegraph square," the protected area around which the rudimentary observatory buildings had been erected. A flag can be seen extended in the wind.

This is an image of observation turned back on itself, a bold reminder to the Royal Society of just who had made the journey to

FIG. 3.8. The Alta Vista Observatory, where a series of rough buildings formed a "telescope square," in a stereo-photograph by Charles Piazzi Smyth. Credit: Royal Observatory Edinburgh.

the top of the mountain, and what he had accomplished there. If the purpose of the expedition had been to subtract the atmosphere from celestial observations, Piazzi Smyth had surely accomplished that. He'd also shown that every observation of even the most distant celestial objects was also an observation of the earth's atmosphere. Finally, he'd understood that every outward observation was also, inevitably, an inwardly turned observation of the observer himself, of the self at the telescope's eye.

Everyone agreed that the expedition had been successful in proving the merit of mountaintop astronomy. Still, Piazzi Smyth managed somehow to wring defeat from the jaws of victory. A group of referees contacted by the Royal Society to judge the work before publication considered that Piazzi Smyth had strayed too far from his area of expertise in his geological and botanical observations, and refused to print the photographs which he had gone to such effort to take, citing cost of publication. Piazzi Smyth responded with a characteristic mixture of petulance and defiance. Within months he and Jessie had published their own account, including all of their photographs and observations. (Piazzi Smyth noted acerbically that his wife had managed to single-handedly print all three hundred photographs needed for the publication.) It was the first indication of what would prove a persistent problem for Piazzi Smyth—transgressing disciplinary boundaries in scientific circles and ruffling scientific feathers.

Piazzi Smyth's restless mind made it hard for him to sit still, and within a few years of his return, he had found a new fascination that would prove even more troublesome. No longer was he fixated on proving the feasibility of observing the stars on top of a mountain. It was to another kind of mountain, a man-made one, to which he now turned his interest. Still fascinated with matters of visibility, the question he now posed was: Is it possible to see God if one looks hard—and measures carefully—enough?

The man-made mountain was the great Pyramid at Giza. It had long been the subject of polite European curiosity. Ever since Napoleon had visited at the end of the eighteenth century, Europeans had wondered how the pyramids had been built and by whom. The ratio between the perimeter of the base of the pyramid and its height, for example, was the same as that between the circumference of a circle and its radius, suggesting, to those who wished to believe it, that the ancient builders had understood pi. More intriguingly, but more complicatedly, in the 1850s a man called John Taylor suggested that the basic unit of construction of the pyramid was a cubit of twenty-five British inches long. The British inch, according to Taylor's assessment, had an ancient pedigree. Not only that, Taylor inferred that this ancient British inch was also a sacred British inch, having formed the basis for the cubit which Noah had used to build his Ark and Moses his Tabernacle.

Piazzi Smyth read Taylor's work and was so taken with his ideas that he applied his considerable writing skills to transforming Taylor's obscure pamphlet into an exciting narrative dramatizing the Pyramid's divine origins and the correspondingly divine lineage of the British inch. His book *Our Inheritance in the Great Pyramid* was the product of just six months of intense work, but it immediately found a large and enthusiastic audience of readers.[27] In the light of

FIG. 3.9. Charles Piazzi Smyth wearing an Egyptian fez. Credit: Royal Observatory Edinburgh.

competition with the French over the metric system, many were happy to look at the Pyramid along with Piazzi Smyth and see evidence for the divinity and antiquity of British metrical values. Soon, he and Jessie embarked for a self-financed journey to visit the Pyramids and measure them for themselves. If anyone could look hard enough and see well enough to find evidence of divinity stamped upon the stones, it was Piazzi Smyth.

The result of Charles and Jessie Piazzi Smyth's stay at the Pyramid was thousands of measurements, made with, among other things, a "well-seasoned" rod from a pipe organ dating from the time of Queen Anne which would be less likely to warp in the intense heat, along with more modern mahogany sliding rods and ivory scales. The Piazzi Smyths observed the Pyramid as carefully as anyone, measuring its dimensions in as many directions and with as much precision as possible. At the same time, they made meteorological and astronomical observations much as they had atop the mountain at Tenerife. Piazzi Smyth proudly presented their results to the Royal Society in April 1866, a year after their return. He was rewarded for his efforts with a prize from the Society recognizing the "energy, self-sacrifice and skill" with which he had undertaken the work.[28] It would seem that Piazzi Smyth had managed to bring the Pyramid into the realm of precise and incisive observations that he had triumphantly entered on top of Tenerife, and in so doing to read God's intervention in the form of the structure. But while the quality of Piazzi Smyth's measurements was never in doubt, his inferences from them ultimately went too far. Though it did not happen immediately, Piazzi Smyth's reputation among his scientific peers was irredeemably damaged by his commitment to the idea of a sacred origin for the British inch.

The matter came to a head some ten years after his trip to the Pyramids, when Piazzi Smyth submitted a paper on the topic to the Royal Society in which he accused the renowned physicist James Clerk Maxwell of "serious error in an Egyptian allusion" Maxwell had made in a lecture to the British Association for the Advancement of Science.[29] Piazzi Smyth's paper was rejected as unsuitable, since it constituted what was viewed as an ad hominem attack on Maxwell. Piazzi Smyth's ill-judged response was to resign his fellowship from the Royal Society. He did not expect his hasty offer to be accepted. Much to his surprise—and chagrin—it was. At the age of fifty-five, Piazzi Smyth had managed to exile himself from the Society that arbitrated the scientific world in which he had lived his entire life.

Though Piazzi Smyth's friends were sympathetic, most felt he had brought the sad state of affairs upon himself. This poignant self-exile from the scientific community was perhaps one reason his next passionate engagement was with an instrument that freed him completely from the need to coordinate, communicate, or calibrate with others. The rainband spectroscope, as it was called, allowed him to do science alone. With it, he could look to the skies as an isolated individual and diagnose the entire contents of the atmosphere. What Piazzi Smyth hoped the spectroscope could do was something far more radical even than freeing him from the restricting embrace of the parliament of science. He hoped it could do nothing less than help transform meteorology into a predictive, rather than a descriptive, science.

Keeping in mind that astronomy had long provided the template for a successful predictive science (even as a new emphasis on physical speculation had crept into it), anyone wishing to predict the weather on scientific lines was faced with the challenge of matching astronomy's predictive power. This was, to put it mildly, not easy. In the 1870s, weather forecasting was within scientific circles possibly even more controversial than mystical theories of the Great Pyramid.

Beginning in 1859, Admiral FitzRoy embarked upon what he called an experiment in weather forecasting. Using a telegraphic network that had been established merely to gather weather data into a system for generating and communicating weather forecasts, FitzRoy fashioned himself as a one-man meteorological band. He based his forecasts on observations of barometric pressure, temperature, and observed wind speed taken at a dozen locations around the country and transmitted to him via telegraph. Applying rules of thumb and his sailor's intuition, within thirty minutes of receiving the information each morning he sent his forecasts back out over the network. The forecasts were immensely popular with local fishermen and sailors, as well as holiday-goers seeking sunshine. They were also controversial because they were so often incorrect. What good was a government science office, critics queried, that sent out erroneous predictions? Surely it did no good for the science of meteorology to be tainted by such inaccurate forecasts. Much to the discomfort of those scientists who winced at every inaccurate forecast, FitzRoy's program received a great deal of attention from pundits and commentators who called him a "weather prophet" and made much of the humorous concatenation of scientific intent with the kind of fairground prognostications made by fortune-tellers. Things came to a sudden halt in 1865, when FitzRoy committed suicide for unknown reasons.

Following FitzRoy's death, a committee formed of fellows of the Royal Society had been appointed to oversee the activity of the Met Office. They were unhappy to discover that the government office had been run as a personal meteorological fiefdom. FitzRoy had delegated little, and written down less. He used no scientific laws or mathematical equations, relying on his sailor's intuition to produce forecasts by himself that he saw as augmenting rather than supplanting the weather wisdom of self-sufficient sailors. The Royal Committee members disapproved of what they considered government sponsorship of an act of individual prognostication akin to fortune-telling. Fearing liability for deaths at sea should the warnings be incorrect, and concerned to protect the reputation of the infant science of meteorology from charges of amateurism, they shut the storm warning program down.

Ten years later, the project of government-sponsored storm warnings in Britain remained locked in a stalemate. The fishermen and sailors along the coast missed the forecasts and wanted them reinstated. The committee of scientists still resisted, suggesting instead a round of internal, private forecasts. In the meantime, the *Times* had decided to go ahead and publish a weather map, the first of its kind, in the daily newspaper, beginning on April 1, 1875. Piazzi Smyth, for one, saw the resemblance between the government storm warnings and the folksy weather wisdom of old, but unlike the Royal Society fellows, it didn't bother him in the least. He despaired of the bureaucracy and what he considered the unnecessary punctiliousness of the Royal Society committee members. He bemoaned those scientists who wanted to keep science for themselves, to claim a higher knowledge of phenomena—the movements of clouds, vapor, heat, and cold—which remained as unruly as the crowds that jostled on railway platforms on their way to seaside recreations. Excited by the novel technology of the rainband spectroscope, he saw an opportunity to circumvent the Royal Society committee and to bring weather forecasting back to the people. He recognized that the spectroscope could also sort out the links between terrestrial and heavenly phenomena at the same time, clarifying the unity of nature as well as the specific properties of earthly weather. Piazzi

Smyth's plan was to use a descendent of the spectroscopes that he had carried up the mountain at Tenerife, and deployed in a variety of exotic locations ever since; it would enable him to diagnose perhaps the most changeable, fluctatory phenomena on Earth—the skies above Britain. The weather embodied a deep paradox. It was made up of uniform molecules, and yet it was eternally in flux.

Spectroscopy had developed rapidly in the years following his Tenerife expedition. In 1859 Gustav Kirchhoff and Robert Bunsen had shown that the lines in the solar spectrum corresponded to chemical contents of the atmosphere of the sun, and Kirchhoff had gone further to correlate many of the Fraunhofer lines with specific metals. But there was still much confusion over what exactly caused the lines, whether some were the result of absorption in the sun's atmosphere, some of absorption in the earth's atmosphere, and some, possibly, owed their presence to a substance present in both. In 1860, David Brewster published a long paper, coauthored with J. H. Gladstone, in which he "majestically mapped the separation" between solar and terrestrial lines. The capstone of nearly three decades of work was the publication of a five-foot-long map of the solar spectrum in which he clearly distinguished between solar and atmospheric lines (without making any guess as to what might cause the lines). In it, Brewster and Gladstone referred to Piazzi Smyth's Tenerife observations, noting that he "had an opportunity of analyzing the light of the sun when seen through a smaller amount of atmosphere than has fallen to the lot of any other investigator."[30] Despite their success in mapping the lines, an experiment designed by Brewster and Gladstone to reproduce the bands in the laboratory had failed and the origin of the atmospheric lines remained unexplained.

It took a Frenchman to clarify the matter. In 1865, Jules Janssen had stood on the balcony of his house on rue Labat in Montmartre, Paris, and aimed a spectroscope at the sky. Janssen was poor, and the earth's atmosphere was a ready and free laboratory. Following up on the curious phenomena already identified by Brewster in 1833, he wanted to investigate the same lines that Piazzi Smyth had seen on the Tenerife mountain, and to try to pinpoint their origins to the earth or the sun's atmosphere. Using a very good prism, he could see something that no one else had seen—the so-called dark bands were in fact crowds of dark lines, similar in structure to the more familiar, less variable lines of the spectrum that had been initially identified by Fraunhofer, some of which had been determined to be solar in origin. Janssen observed the lines at all times of day and noticed something further. The lines were especially strong when he observed sunlight at sunrise or sunset, but they never disappeared, even when he looked at the sun at high noon (a finding that contradicted Brewster's earlier work). They must, he reasoned, be caused by something ever present in the earth's atmosphere. (The effect at sunrise and sunset would be greater because sunlight had to pass through more of the earth's atmosphere to reach him.)

He set out to figure out what it might be. First he traveled to a Swiss mountaintop, to see if the lines were diminished when viewed through a smaller portion of atmosphere. They were. Then he made his way to Lake Geneva, where he observed a large bonfire on the pier at Nyon. When viewed from close by through the spectroscope, he saw no dark bands, only the normal spectrum. But when viewed from the top of the tower in Geneva, a dozen miles across the lake, the spectrum was crossed by the same dark lines that Piazzi Smyth had observed on Tenerife and that Janssen had observed over Paris. He was by now almost certain that they were caused by water vapor that was suspended in the air above Lake Geneva. With the cooperation of a Parisian gasworks, he set about making an artificial atmosphere to confirm his guess. A span of metal tubing stood in for the breadth of the entire atmosphere. Buried in a box of sawdust and enclosed at either end with panes of glass, the tube was filled with water vapor under pressure. At one end of the tube, an array of gas jets sent a beam of strong light through the tube, while Janssen observed at the other end of the tube with his spectroscope. He saw the same dark bands he'd seen over Paris, which had first been noticed by Brewster in 1833 and which Piazzi Smyth had seen wax and wane on top of the mountain in Tenerife in 1856. The more the pressure was raised, the darker the bands appeared. The greater the length of tubing, the darker they appeared. Janssen was now certain that the dark bands were caused by water vapor in the earth's atmosphere. He wasted no time in concluding that the lines could also be used to search for water vapor in the atmospheres of other celestial objects. He asserted immediately, for example, that there was no water vapor present in the atmosphere of the sun, a remarkably assured statement.

Piazzi Smyth's own interest in spectroscopy, which had waned after his Tenerife expedition, was reignited by the news that Janssen had managed to observe the solar prominences for the first time during a total solar eclipse in India in 1868. Soon after the Indian eclipse, Piazzi Smyth bought himself a new spectroscope. It was a small wooden device just four inches long and less than an inch in diameter, with an eyepiece at one end and a diffraction slit at the other. Inside lay a series of alternating prisms, cleverly arranged so that the light that passed through them emerged from the spectroscope at the same angle at which it had entered. While not a poor man's device (costing some two pounds), the pocket spectroscope was the province solely of neither professional meteorologists nor astronomers.

The instrument made it possible to do solar physics at a moment's warning, anywhere the fervent observer happened to be, liberating him or her from surveillance. Skill was required, a delicate habit of minute adjustments, to get good results. The instrument should be aimed low down near the horizon so as to subtend as thick a portion of the atmosphere as possible. A break in the clouds was ideal, but the sun should not interfere too directly, lest it overpower the delicate observation. Sunrise and sunset were also not recommended for this reason. Fogs, mist, and dense coal smoke could also obscure the readings.

For Piazzi Smyth, all this adjusting was enjoyable, part of the fun of looking at the skies. The convenient little tool found its way to his eye many times a day. Fifty times was easily possible when he had his "enthusiasm-fit" upon him.[31] Every situation was different, every cloud break, every confluence of pressure, temperature, and wind bearing variously on the transformations of the weather. He didn't know exactly what, if anything, he was looking for. He didn't really need to be looking for anything, at any particular moment. He looked first, almost before thinking.

In 1875, Piazzi Smyth made a trip to Paris, where he visited the Astronomer Royal Urbain Leverrier (he found him exceedingly rude, leaving him and Jessie to find their way home alone through a thunderstorm). That storm followed the Piazzi Smyths back to London and, observing it carefully with his spectroscope, Piazzi Smyth noticed a hazy and indistinct—but nevertheless present— "dark broad band" crossing the spectrum between the red and orange sections. The band was darker than the areas surrounding it, and was fuzzy rather than distinct. It faded as he moved the spectroscope to another part of the horizon. When he took his eye away from the instrument and looked with plain sight at the sky, he could see nothing in particular that looked different in that region than in any other. Piazzi Smyth continued to observe as he traveled north to York, where he noticed the strong presence of a blurry dark band in the spectroscope one sunny morning when rain seemed unlikely. Rain did indeed ensue, and observing it fall, Piazzi Smyth felt vindicated in his hunch that the pocket spectroscope was a special new kind of tool for predicting the weather.

He set out to share the news of what he'd come to call the "rainband." Unlike the hard-edged lines of the solar spectrum, which were understood to be caused by the absorption of light waves by different substances in the solar atmosphere, the rainband was blurry, indistinct, and dynamic. Just what caused it, Piazzi Smyth did not say, though his identification of it with the coming of rain was a strong indication that he thought it most likely to be the signature of water vapor in the atmosphere. He announced his discovery in a letter to the journal *Nature*, founded just months earlier, under the title "Spectroscopic *prevision* of rain with a high barometer." The title made it clear that Piazzi Smyth was modifying what had previously stood as the most basic of meteorological assumptions— that a high barometer, and therefore high pressure, implied good weather.

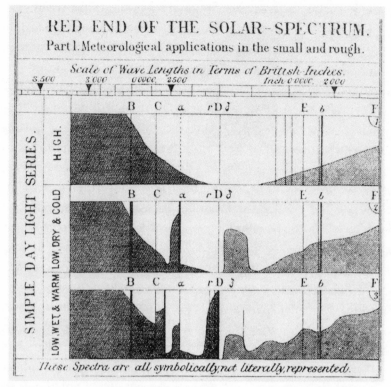

FIG. 3.10. Solar spectra recorded under different weather conditions showing the position of the rainband at r.

In fact, Piazzi Smyth had revealed precious little about the rainband. He did not offer an opinion on what caused it, and did not even mention the phrase *water vapor* in his article. More than a new bit of scientific knowledge, what Piazzi Smyth was so eager to announce was a new tool for doing science. Spectroscopy—recently identified as a tool for diagnosing the contents of the earth's atmosphere—could also be a device for measuring the *changing* amounts of those elements in the earth's atmosphere. This made it a tool of practical meteorology, what he called the "prediction of weather for the common purposes of life."

Despite Piazzi Smyth's excitement, there was reason to be very cautious about the possibility that the "rainband" spectroscope might transform weather prediction. The blurry bands of the so-called rainband were even harder to learn to see than the fixed solar lines, since they were variable. They changed because the thing they represented—water vapor in the atmosphere—was itself constantly shifting. The allegedly convenient and easy-to-use rainband spectroscope was in fact an instrument for diagnosing variability of a very complex and very particular sort. While spectra contained an enormous amount of information about the entire breadth of the atmosphere at a glance, they were also snapshots, representing just an instant in time. To be useful, they had to be interpreted in relation to snapshots that came before and those that came later. Even the darkness of the bands was only useful when compared with preceding rainbands. Comparing the intensity of successive rainbands was an intensely subjective task, which only relatively few could master sufficiently to render it a consistent practice. F. W. Cory, writing in the *Quarterly Journal of the Royal Meteorological Society*, argued that the spectroscope was simply too difficult to use, requiring two or three months of "patience and perseverance" (this admitted by one of its most ardent supporters) to be mastered easily by unsupervised individuals.[32]

Piazzi Smyth persevered. He wrote a series of letters to popular journals in which he tried to reframe the problem of prediction within meteorology, a problem that had always been knotty but had become particularly so ever since FitzRoy's sensational forecasts and their cancellation following his sensational death. For Piazzi Smyth, prognostication was not a bad word, and the pocket spectroscope no better or worse than existing methods for predicting the weather. It was merely one more tool for energetic, self-motivated people who couldn't wait for a "perfect" meteorology (akin to astronomy) to rise from the ashes of Admiral FitzRoy's storm warning project. "We need not after all be offended at the mere name of 'prognostic,'" Piazzi Smyth reassured the public, pointing out that even the stalwart barometer could only go so far in predicting the coming weather. The boundary between folk wisdom and science, which the Royal Society committee was desperate to police, was a mere illusion, according to Piazzi Smyth. Knowledge was knowledge, however it was gained, and in relation to as complex a thing as the weather remained necessarily provisional, dependent on skill, judgment, and the individual perspective of a single observer whether it was generated by the most rustic fisherman or the most well-trained astronomical observer. "For are there not prognostics and prognostics in meteorology! What are not the risings and fallings of the wind-compelling barometer itself, but a weather prognostic for those who can interpret them."[33]

In the face of the almost unimaginable complexity of the weather, Piazzi Smyth was both pragmatic and optimistic. Rather than limiting the sources of data or its applications, he thought it made sense to increase both. He imagined legions of independent observers across the country, "many, very many people," the natural acuity of whose eyes was enhanced by the pocket rainband spectroscope to render them an army of proto-Supermen, able to see through the clearest of skies to the vapor that lurked within. Thus outfitted, they could "observe and speculate on the weather for themselves at their own places of abode, supplementary to any forecasts that may be issued once a day from London." The best part about the spectroscope was its portability and ease of use. It enabled any individual to penetrate his solitude and immobility to reach, quite literally, beyond the confines of terrestrial domesticity and into the vastness of space itself. A glance of just two seconds was enough to "tell an experienced observer the general condition of the whole atmosphere."[34] It gave a feeling of security, Piazzi Smyth explained, to know that even in the most cramped and confined space, "with no more than a few cubic feet of peculiar, and for science-purposes, vitiated, air about it," he could nonetheless still be "nobly looking through the whole atmosphere from the surface of the earth right through to space outside, and analyzing its condition as to watery vapour (the raw material of rain, as the *Times* phrased it) in one instantaneous, integrating glance."[35]

The information the spectroscope gave to the observer was not merely fast but all-encompassing. It penetrated the entire atmosphere at the speed of light and made a single human being the diagnostician of the globe. This was a global science of the atmosphere practiced by individuals. Unsupervised, untethered, unabashedly personal—the pocket rainband spectroscope fulfilled Piazzi Smyth's

fantasy of how science should be. Here was a tool with practical benefit to the sailors, farmers, and holidaymakers of Britain and beyond,
a tool which could deliver immediately on its promise, rather than
coyly holding out the hope for future understanding of the laws
which drove the weather. With it, every man could become an
observatory, twitching the skies above.

But what Piazzi Smyth considered the very best qualities of the
spectroscope—the way it facilitated a multitude of quick and calibrating glances, its constant availability to those who found themselves gripped by it—proved to be its undoing. Instead of bringing meteorological observing to the people, as Piazzi Smyth had
hoped it would, the miniaturized, portable rainband spectroscope
revealed just how ill suited most people were to the practice of
science. The craze for rainband spectroscopy faded as quickly as a
summer storm, leaving the public to rely on their own eyes, their
familiar barometers, and the unfamiliar new weather maps when
it came to making decisions about the weather.

While Piazzi Smyth was promoting the benefits of individualized
vision, the scientific winds were blowing the other way. A Parliamentary investigation into the Met Office was begun in October
1876. In the wake of FitzRoy's death, the Met Office had embraced
so-called self-registering instruments that could automatically
trace the changes in the weather. These fantastical devices were like
instrumental chimeras, combining the normally separate actions of
measurement and registration in one object. The bedeviling influence of friction, which had foiled attempts to design self-registering
instruments in the past, met its match in the form of photography.
One of the first applications of photography was to the challenge
of automatically registering the weather. In 1845, just six years
after Louis Daguerre had pioneered the photographic process, two
meteorologists (Francis Ronalds at Kew Observatory and Charles
Brooke at Greenwich) set about designing a series of self-registering
instruments (including a magnetometer, electrometer, barometer,
and thermometer) which could deflect a beam of light against a photographic place. Other self-registering instruments used the simpler
method of connecting an inked pen to the measurement device.[36]

These traces were to be used not for weather forecasts—which FitzRoy's death had revealed as dangerously subjective—but for the long-term project of deducing the physical laws that underpinned the movements of the atmosphere. In the eyes of Robert Scott, FitzRoy's replacement as head of the Met Office, the values of comparison and continuity completely trumped those of independence and skill. In 1875, he quoted approvingly from the 1840 report of the Committee of Physics and Meteorology of the Royal Society: "Systematic cooperation is the essential point to which at present everything else must be sacrificed; and cooperation on almost any plan would most certainly be followed by more beneficial results than any number of independent observations, however perfect they might be in themselves." The contributions of those who the committee referred to as "amateurs of science" were welcomed only as long as they conformed to the rules, "even," it was noted, "at the temporary sacrifice of their own views and convenience."[37]

Exactly how the ever more numerous traces of pressure, temperature, and other weather-related phenomena could be transformed into a science of the weather was an open question. British meteorologists in the 1870s felt themselves to be stalled in a stage of early development. What astronomers had once achieved—the ability to make predictions that were accurate far into the future—seemed an ever-receding goal. Meanwhile, astronomy itself had forfeited the self-confidence it had assumed in the wake of Newton's great achievement. Men such as William Herschel, Edward Sabine, John Herschel, David Brewster, Jules Janssen and, last but not least, Charles Piazzi Smyth had shown that astronomy could be a physical as well as a positional science, but in so doing they had exposed new sources of ignorance and uncertainty. The mature science of astronomy was young again, while meteorology, ever the "infant" science, sought new sources of confidence.

The scandal of FitzRoy's death did not help matters in Britain. The promise of automatic observatories was a decidedly mixed one—offering the chance of fulfilling Humboldt's dream of disaggregating the signals of a pluripotent Cosmos but challenging the ability of scientists to manage ever greater amounts of data. Data had long threatened to overwhelm singular astronomers as they faithfully recorded the skies in a single location, night after night. Once armies of self-registering instruments were unloosed on the observatories of the globe, it was hard to see how it would ever be possible to catch up.

New techniques for reducing the traces of such instruments were urgently required. Humboldt had understood this back in the 1830s and had urged Heinrich Berghaus to publish a graphical companion to his *Cosmos* in the form of a *Physikalischer Atlas*, which used diagrams to represent the way climate, plants, animals, and geological features changed across the globe. In Britain, Francis Galton came up with a strikingly visual way of finding the mean values for meteorological traces that involved superimposing a series of traces and graphically determining the average line. As innovative as these visual methods were, they could only go so far in the absence of data. In exchange for the personal judgments of forecasters, self-registering instruments promised objective knowledge and produced reams of data. How to wring meaningful understanding from the proliferating traces of the atmosphere was less than obvious. Not all data was created equal. One could have simultaneously too much data of one kind and not enough of another.[38]

One problem was that the atmosphere was three-dimensional but observers had been largely limited to data from the surface of the earth. This was part of the reason that Piazzi Smyth had been so taken with the spectroscope. It enabled anyone who used it to soar high overhead, traversing unimaginable distances in the process. This was its great advantage, but it was also a disadvantage, as the spectroscope was unable to distinguish the absorptive capacities of different parts of the atmosphere. Every molecule that lay in the line of vision was included in its gaze. It flattened the heterogeneity of the atmosphere even as it provided a way to diagnose its changes. This was a paradox that Piazzi Smyth, for one, was willing to accept, considering the benefits of this kind of vision accumulated to outweigh the losses.

But there were many ways to see the skies. One of the challenges was to get up into the atmosphere and observe it *in situ*. From the 1850s onward, a series of daring and popular balloon journeys up into the atmosphere took place. These sensational flights were undertaken in the service both of meteorological knowledge and with the spirit of adventure that characterized polar expeditions. They were, as far as they went, immensely successful in raising the profile (literally) of meteorology and capturing the imagination of the public. But balloons were expensive, and the journeys dangerous. At best, they provided a single set of observations recording the conditions in one particular column of air over the course of several hours. Generating systematic knowledge out of these singular adventures would be almost impossible.

Another way to get up into the skies, figuratively, rather than literally, was to pay close attention to clouds. Clouds rode the air currents that determined weather and climate. Noting their sizes, their shapes, and their movements was a way to map the invisible ocean of air above. Clouds were like flags in the upper atmosphere, telling an observer which way the wind was blowing and in which direction, as well as giving an indication of how much water vapor was present in the air. If that were the case, then clouds could be a way to see deeper into the mysteries of the atmosphere. Rather than obscuring the heavens, or frustrating astronomers, clouds could reveal the patterns of atmospheric movement and the laws that drove air around the planet.[39]

The first step was to classify the clouds. When Luke Howard had introduced his pioneering cloud nomenclature at the beginning of the nineteenth century, he had helped regularize the study of clouds. What he had not been able to say with confidence was whether his tripartite system of clouds would apply throughout the world. Were clouds globally uniform, or were certain clouds only to be found in certain parts of the world? The century had almost ended before anyone had seriously attempted to answer this question. In 1885, an amateur meteorologist with deep pockets named Ralph Abercromby decided to try. He set out on a self-financed circumnavigation of the globe with the explicit aim of determining how universally applicable Howard's system was. He determined that while the same type of cloud could signify different kinds of coming weather in different places, the essential cloud types were indeed universal.[40]

Abercromby described his clouds with words, but his discovery inspired the search for more methods of faithfully capturing clouds. If clouds were universal, they could unlock the mysteries not only of the local weather but of what it seemed increasingly reasonable to assume were planetary weather patterns. But in contrast to temperature, pressure, or even rainfall, clouds resisted instrumental registration—they belonged to that "extensive class of phenomena which cannot be recorded instrumentally, but of which it is necessary to take careful notice owing to their importance as indicating changes which are in progress in the atmosphere." Clouds were almost impossible to observe with the kind of objectivity that instruments promised, but they were too important to ignore. "It is very difficult," noted C. H. Ley in the preface to his father's contribution to cloud classification, *Cloudland,* "to treat of a vague and complicated subject in any but a vague and complicated manner."[41] What was needed was an instrument that could register the clouds in the same way that the barometer registered pressure and the thermometer temperature—instantly, faithfully, and reliably. By the 1870s, exposure times were fast enough that the photographic camera presented itself as a potential solution to this problem.

It comes as no surprise, given Piazzi Smyth's lifelong commitment to watching the skies, that he was an early advocate for what he called cloud-capturing photography, a new technology to feed his endlessly voracious visual appetite. He'd grown up learning to use whatever tool suited the task of observation before him: Watercolors, pen and ink, pencil, and paints were his early tools. As a youth in the Cape of Good Hope, he had also recognized that photography had the potential to transform scientific observation. Throughout his life, he'd experimented with it, making photographs on board ships, atop mountains, and in the gloom of Egyptian pyramid tombs, developing stereo-photography and even the photography of plaster models. In the 1870s, he designed a new camera specifically for taking photographs of clouds. It incorporated a special corrector to counteract the spherical aberration otherwise introduced by a portrait lens, enabling the full aperture of the lens to be used without distortion.[42] He exhibited it at the Edinburgh Photographic Society's 1876 Exhibition, alongside some cloud photographs, and was awarded a silver medal for it.

He then abandoned the project to undertake one last, intense piece of spectroscopic research, seeking clear rather than cloudy skies in which to probe the nature of the solar spectrum as deeply as possible. Instead of Tenerife, he traveled to the more accessible Portugal and there found that he was able to almost eliminate the so-called telluric, or earthly, lines associated with water vapor. There Piazzi Smyth was finally able to separate the true solar spectrum from both the dry atmospheric spectral lines and those corresponding to moisture in the atmosphere, the culmination of the project he'd begun atop Tenerife some twenty years earlier. As proud as he may have been of the fruits of his observational labor, Piazzi Smyth was not alone in his quest to subtract earthly from solar phenomena.[43] At the 1882 meeting of the British Association for the Advancement of Science, several others claimed priority in the matter of separating the solar from the dry and wet atmospheric lines. The related question of whether the oxygen bands had their origins partly in the sun also remained unanswered well into the 1890s. In 1893, an elderly Jules Janssen decided to try to settle the matter himself. At age sixty-nine, he made his way to the top of Mont Blanc in an attempt to observe solar oxygen from there. His observations of the absence of oxygen lines in the solar spectrum as viewed from the top of the mountain were taken as evidence for the absence of oxygen on the sun. The spectroscope continued to amaze with its ability to penetrate deep into the atmospheres of distant objects.

While the scientific community continued to seek evidence for new substances in the atmospheres of the earth, the sun, the planets, and even such remote phenomena as the Zodiacal light, Piazzi Smyth himself receded ever further from public view. His last undertaking was almost entirely solitary. He had both embraced the pleasures of independence and tasted the bitterness of exclusion during his lifetime. His commitment to pyramidology had resulted in his

FIG. 3.11. Charles Piazzi Smyth in old age with a grand-niece. Credit: Royal Observatory Edinburgh.

self-imposed retreat from the Royal Society and the community of scientists it represented. His spectroscopic work continued to be of excellent quality, but his insistence on using the British inch as the unit of length had severely limited the usefulness of his maps. In the end, he found himself almost alone. He had time, the luxury of retirement. He had his instruments and a few assistants to help him. And he had the clouds. They passed by the high windows of his home in Ripon, Yorkshire. He grasped them with his camera, angling it up to exclude all but the tops of the tallest trees.

He returned to cloud photography, the subject he had investigated some twenty years earlier, looking for a way to standardize the observation of the clouds just as the spectroscope had standardized the observation of light. But while his earlier project had been undertaken in the acknowledgment of a wider community of observers, this final project, undertaken in 1892 and 1893, was an impossible, almost lunatic attempt to record the face of the skies

FIG. 3.12. Photograph from *Cloud Forms That Have Been at Clova, Ripon*, taken from his library window by Charles Piazzi Smyth on June 30, 1892. Credit: Royal Society.

alone. He himself gently mocked his project, calling it a "labour of love and meteorologic research, in days of old age and failing faculties." At the heart of his project was a set of photographs which, without intention or ambition to extend beyond their tiny window of the world, served as a powerful renunciation of the communal approach to knowledge. Of the hundreds of images he took, he printed 144 of the best in three massive volumes, bound in leather and prefaced with a manuscript copied out in a clear hand, detailing the nature of his project. These tomes represented the work of thousands of hours, but they were read by almost no one. Never published or widely shared, they stand as a monument to life spent in intensely personal observation.[44]

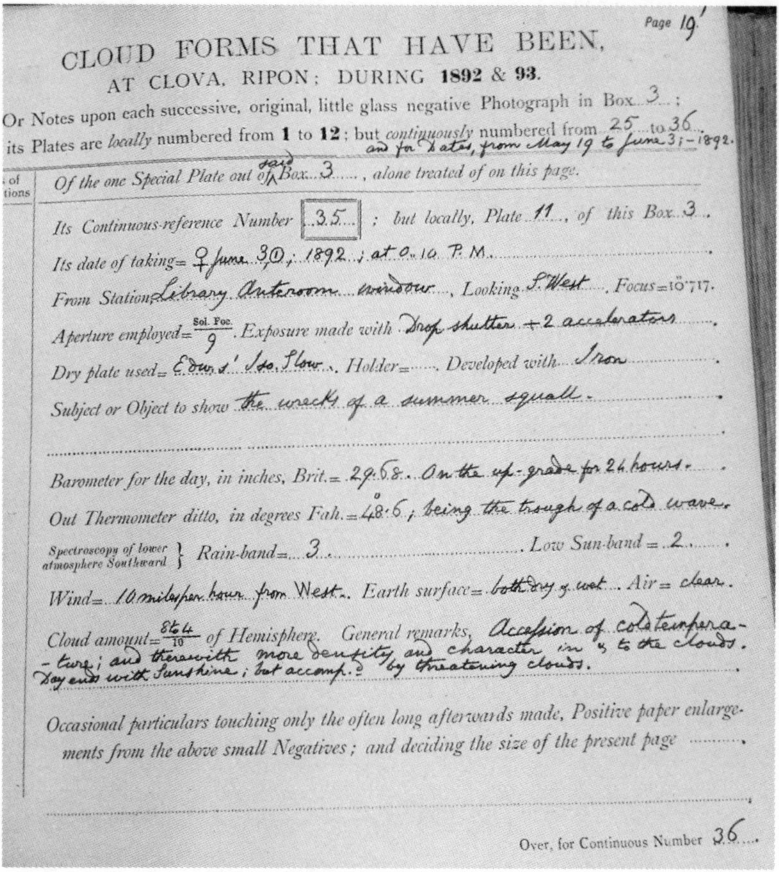
Page 19

CLOUD FORMS THAT HAVE BEEN,
AT CLOVA, RIPON; DURING 1892 & 93.
Or Notes upon each successive, original, little glass negative Photograph in Box 3 ;
its Plates are *locally* numbered from 1 to 12; but *continuously* numbered from 25 to 36
and for dates, from May 19 to June 3; – 1892.

Of the one Special Plate out of said Box 3, alone treated of on this page.

Its Continuous-reference Number 35 ; *but locally, Plate* 11, *of this Box* 3.
Its date of taking = ♀ June 30; 1892; at 0.10 P.M.
From Station Library Anteroom window. *Looking* S. West. *Focus* = 10.717.
Aperture employed = Sol. Foc. / 9. *Exposure made with* Drop shutter + 2 accelerators.
Dry plate used = Edwards' Iso. Slow. *Holder* = . *Developed with* Iron
Subject or Object to show the wrecks of a summer squall.

Barometer for the day, in inches, Brit. = 29.68. On the up-grade for 24 hours.
Out Thermometer ditto, in degrees Fah. = 48.6; being the trough of a cold wave.
Spectroscopy of lower atmosphere Southward } *Rain-band* = 3 . *Low Sun-band* = 2.
Wind = 10 miles per hour from West. *Earth surface* = both dry & wet. *Air* = clear.
Cloud amount = 8 to 4 / 10 *of Hemisphere. General remarks,* Accession of cold tempera- ture; and therewith more density and character in ⁓ to the clouds. Day ends with sunshine; but accomp.d by threatening clouds.

Occasional particulars touching only the often long afterwards made, Positive paper enlargements from the above small Negatives; and deciding the size of the present page

Over, for Continuous Number 36.

FIG. 3.13. "The wrecks of a summer squall," Smyth notes on the cloud photograph taken on June 30, 1892, along with observations of barometric pressure, temperature, rain- and sunband, and wind speed. Credit: Royal Society.

Piazzi Smyth's timing was, in a certain sense, excellent. At the very moment that he was devoting himself in solitude to his "labour of love," the scientific community had begun its own project in cloud photography. Inspired in part by Abercromby's discovery that cloud types were universal, in 1891 the attendees of an international conference of thirty-one meteorological directors from around the world launched an international project to map the clouds. Their plan encompassed not merely a single location but, in theory (if not in practice), the entire globe. The projected International Cloud Atlas would improve upon Howard's classification system and put flesh on Abercromby's anecdotal assertions about the universality of clouds. Headed by a Swede, Hugo Hildebrandsson, and a French-man, Léon Teisserenc de Bort, the atlas had the crisp mark of impe-rial power on it, the confidence to mark the skies with order as the railways and telegraph wires had marked the land. The plan was explicitly both global and synchronized, aiming to promote what its authors described as "inquiries into the forms and motions of clouds by means of concerted observations at the various institutes and observatories of the globe."[45] It was, in other words, everything that Piazzi Smyth's solitary project was not.

Clouds became, through the classificatory magic of the Interna-tional Cloud Atlas, standardized objects that could be identified reli-ably on the basis of images that served much the same purpose as the sketches of birds in a naturalist's guide. Clouds were, the atlas declared, universal types that could be identified at any point on the planet. To aid in such identification, the atlas made stunning use of color photographs. It proved impossible, however, to cap-ture images of all the sixteen basic types of clouds described in the atlas this way. Certain clouds, such as the alto-stratus, the nimbus, and the stratus, were too difficult to catch. Lithographic represen-tations of painted pictures—an older technology for representing ideal types—of these cloud types appear in the completed atlas alongside color photographs, as do black-and-white photographic images. Perfect scientific vision, these different kinds of images seem to suggest, was impossible. Instead, the act of looking was an active and dynamic process. Seeing clouds well—which included seeing them as universal types—required seeing them in different ways and from different locations. This way of looking proved resil-ient. The International Cloud Atlas has remained in print ever since. Photography remains the standard technology for representing the clouds and, just as important, the atlas is a compendium of global knowledge, generated by observers positioned all over the planet just as Hildebrandsson and Teisserenc de Bort had suggested back in 1896. The projected future of meteorology that they envisioned has, in many important respects, come to pass.

In another, and perhaps more important sense, meteorology has changed dramatically since then. No longer is taxonomy enough.

Once a system for classification had been established—not by a solitary looker like Piazzi Smyth but by a committee representing the international community—the next step was to apply it to the deeper, and much knottier, problem of explaining why the clouds appeared where they did, of explaining what drove the weather and the clouds with it. For all its confidence, the Cloud Atlas was little more than a down payment on a future piece of work whose success was far from certain.

Piazzi Smyth offered no help with that project either. His life and his considerable energies had been expended in the belief that looking was an end in itself, an activity that was both morally and scientifically productive. Looking at things that were difficult to see—the distant stars, the ever-changing spectrum, the clouds—was a way to exercise mental faculties and spiritual sense at the same time. Piazzi Smyth left to others the task of seeing through clouds to the physics of the solar system or looking at a fuzzy, shifting rainband to crack the code of the weather, or capturing clouds that shifted with dizzying rapidity. He was ready to look not for explanation but for something deeper still—the mark of the divine. "Up to this time, whatever science can or cannot say in scholastic explanation," Piazzi Smyth reasoned, "or however far behind she may be in reducing either the minute beauties of calm summer skies or the majestic agglomerations of threatening thunder . . . to nothing but a few mechanical processes throughout their whole extent and bearing, yet the forms of beauty exhibited so frequently and prodigally before our neglectful eyes in Clouds, can only be reverentially looked upon by us." In the final consideration, clouds were noteworthy to Piazzi Smyth not because they made elusive, difficult objects for scientific study, but because they made it easy to see God. Able to be witnessed by all of us, they were testaments to divine order, bearing the "visible impress of the greater invisible Intelligence which arranges all we see."[46] At the end of a busy and tumultuous life, Piazzi Smyth took a full measure of solace from the order he found in the wildness of the sky.

4

NUMBER OF THE MONSOON

Arriving in India in 1903 at the age of thirty-five, Gilbert Walker traveled directly to the foothills of the Himalayas. His destination was Simla, the summer capital of the British Empire in India, and the year-round location of the Meteorological Department. He was on his way to assuming the most important job he would ever hold, Director-General of Meteorological Observatories.[1]

He had spent most of the past decade and a half at Trinity College, Cambridge, living among other single men in rooms set off private courtyards. There, he had devoted himself to the study of mathematics. Though he was about to take over the world's most extensive meteorological network, he knew next to nothing about the weather, still less about managing an organization that included hundreds of observatories and tens of thousands of observers. That, in a sense, was the very reason he'd be summoned to India. Things had gotten so desperate, and the problem facing those who hired him so intractable, that Walker's ignorance and lack of experience had started to seem almost desirable. When little else had worked, perhaps it was time to try something completely new. There was one single, highly reassuring fact about Walker: He was one of the best mathematicians of his generation. That was enough reason for Walker's predecessor, John Eliot, to stake his own reputation on the tall, thin young man who had undertaken the long journey to India.

Simla, the city where Walker was headed, was just as paradoxical

FIG. 4.1. Simla and surrounding mountains in the Himalayas. Credit: Wellcome Collection.

as Walker's appointment. It was a cool place in a hot land, a small city in a wildly populous nation, a remote enclave from which to rule a voluminous empire. Located just over 7,000 feet above sea level and nearly 1,000 miles from Calcutta, Simla offered relief from the stupefying heat of the plains. The city had proved its worth by providing a safe, comfortable, and healthy location from which the British could rule their largest colony during the oppressive summer months. Arrayed on a steeply terraced hillside, and governed by the rhythms of the court, the town was as much redoubt as refuge, deriving its power from its very remoteness and exclusivity. "Here," in the words of one visual atlas of the Empire, "wars are planned, peace is made, famines fought."[2]

That last, alliterative phrase was the reason Walker had been summoned.

That remoteness could signal power—that power could be enacted from a distance and that its effects might be amplified rather than attenuated by that distance—was a thing that the British in India already knew. What they didn't know, and what they hoped Walker would tell them, was what powers gave rise to the monsoon deluges

FIG. 4.2. Portrait of Gilbert Walker, who arrived in India at the height of famine hoping to identify the causes of the monsoon.

upon which the well-being—indeed the survival—of tens of millions of Indians and the wealth of the Empire depended.

The idea that mathematics might be able to cut to the heart of a problem was not a new one in 1903. Mathematics had long been granted a special authority among scientists, dating back at least as far as Galileo's measurement of the acceleration of falling bodies. Used properly, numbers could reveal the laws of nature. That mathematics was essential for understanding the atmosphere of the earth had also been long understood, ever since observers had gathered quantitative measurements of temperature and pressure. In the final three decades of the nineteenth century, however, both the mathematical tools, and perhaps even more importantly, the amounts of data to which they could be applied, had increased dramatically. The automatic observatories set up by the Met Office in the wake of FitzRoy's death—recording the traces of the weather day in and day out—were generating data from which it was hoped patterns could emerge. In addition, legions of observers were diligently recording observations by hand. These observations—many of which were made by sailors in the navies and merchant marines of the ruling nations of the globe—had grown exponentially in the decades preceding Walker's arrival in India, for the simple reason that the management of empires depended on the management of the weather. As empires had grown to subtend ever-larger portions of the globe, so the collection of meteorological data had spread to keep pace.

Climatology was the name of the field dedicated to the collection of data about the weather. Originating in German, the term made its way into English and French usage in the first decades of the nineteenth century, rising in usage from the 1840s onward. Its rise can be traced, as so much, back to Alexander von Humboldt's influence at the start of the century (one of the earliest uses of the word was in a French translation of Humboldt's work).[3] Humboldt's concept of climates was related to his insight into the unity of nature. This unity, according to Humboldt, did not produce climatological uniformity. Instead, myriad physical forces combined to generate climatological difference—distinct zones, often defined vertically, in which certain conditions of temperature and precipitation persisted and where certain plants and animals thrived accordingly. If the physical forces of nature were always in flux, the wonder was that they combined in such a way as to produce stable, geographically fixed climates that could be measured, described, and in a certain sense safely stowed away. Safe, in this sense, meant reliable. Humboldt expected them to endure for a long time.

The exploratory and unity-seeking arm of climatology that claims a lineage from Humboldt must be reconciled with the part of climatology that was, as it were, born statistical. In both England and

Prussia, national departments for gathering meteorological data emerged from or were closely linked to state statistical offices. There, the practical benefits of knowing the changing weather patterns—and determining some general climatological rules— were enormous. Understanding climate was of practical benefit in the same way that astronomy, botany, magnetic studies, and sur- veying were—they allowed state holdings to be mapped and under- stood in order that they could be exploited. But perhaps more than any of these allied disciplines, the collection of weather data that could be transformed into climate averages was a nation-building exercise. By gathering data on both citizens and the weather, gov- ernments hoped to control these often-unwieldy phenomena.[4]

Julius von Hann, director of the Central Office for Meteorology and Geomagnetism in Vienna, and Wladimir Köppen, director of the German Marine Observatory in Hamburg, were the primary inheritors of Humboldt's confidence that the earth's face could be measured and known. They differed from Humboldt, however, on the question of how climate should be defined. Armed with the desire, the money, and the institutional capacity, Hann and Köppen transformed Humboldt's mantra of singular exploration into the basis for a systematic discipline. In the process, the Humboldtian emphasis on the interrelationships between living creatures and the environment in defining climate gave way to a definition of climate in terms of fixed zones, identified on the basis of averaged meteoro- logical data. Climate, as defined by Hann and Köppen, was built on the foundations of meteorological data. It was the weather averaged.

More important than the precise definition of climate (as it turned out) was the foundation that these men established for climatology. Under the direction of Hann and Köppen, climatology was a science of rigorous measurement carried out in robust institutional settings. Hann promoted his approach through a *Handbook of Climatology*, published in 1883, which laid out the method by which climatology could advance, step by step, as an invading army. The natural direc- tion for this science was cartographic, and in due course Wladimir Köppen extended climatology by introducing the graphical power of the climate map, on which averaged climatic zones were represented in visually distinctive areas. The "tropical climate" was born on Köppen's maps, alongside the "polar climate," the "subtropical climate," and the "Mediterranean climate."

The success of the program and its weakness lay in the same fact: Knowing the climates of the earth in this way required an avalanche of data. Climatology, in the sense that Hann and then Köppen practiced it, was a science of the telegraph, the postal system, and the publishing house. It relied on an extensive system of measurements for compiling maps, and just as significantly on a system by which such maps could be printed and distributed. This climatological project motivated and absorbed an incredible amount of energy in the decades following its ripest definition.

Despite its productivity, Hann was almost aggressive in setting out the limitations of climatology. He was self-consciously anti-theoretical in his claims for what climatology could do. "Climatology," he cautioned, "is but a part of meteorology when the latter term is used in a broad sense." Climatology, he clarified, was descriptive, while meteorology, which aimed to "explain the various atmospheric phenomena by known physical laws," was theoretical. Nevertheless, the two fields were intimately related. Climatology was an essential part of meteorology, and what it lacked in explanatory power it made up for in its breadth. As a primarily visual science, it provided the means to build up a "mosaic-like picture of the different climates of the work." This was a very orderly patchwork, in which facts were presented systematically. In this way, "order and uniformity are secured, the mutual interactions of the different climates are made clear, and climatology becomes a scientific branch of learning." As Deborah Coen notes, this made climatology a science (at least potentially) of "complex wholes," while meteorology concerned itself instead with reducing atmospheric phenomena to "simpler, theoretically tractable elements."[5]

Much was presumed in the elision here between Hann's admission of the descriptive nature of climatology and his ambitious hope that it could become a true science.[6] How exactly it would be possible to get from the compilation of average values of rainfall and temperature to a science of physical laws was left unclear. There was a deep tension in relation to the nature of change buried in the text of Hann's *Handbook*. Change was essential to this transformation, but just how much change should be defined or investigated remained unclear. Hann included a final section on "Changes of Climate" in his handbook, in which he considered geological changes in climate, such as the ice ages, and the theories of Croll and others who tried to explain them. He also considered the search for shorter, so-called oscillations of climate that could be linked to sunspots. In both cases, however, Hann revealed his overriding commitment to the method of averaging, with its strong assumption of the explanatory validity of a period of stability. To generate an average, as Hann explained, required identifying a period of time across which the average would apply. Using this method, it was eminently possible to identify oscillations either above or below an average value. Change, in other words, was here understood in relation to some fixed period within which averages could be constructed.[7]

Hann's approach was predicated on a commitment to stability in the form of averages, but in practice it enabled him (and others) to identify climatic anomalies. It did so by establishing a rudimentary form of climate system. The statistical table, in this sense, was an inchoate climate model. By bringing together averaged weather data from distant parts of the planet, the table enabled Hann and others to search more easily for patterns within the numbers. Since those numbers corresponded to average weather values—what Hann had defined as climate—they enabled researchers to find connections between longer-term features of the atmosphere. To be sure, the process of averaging elided the dynamism that it was the business of meteorology to uncover and describe—what Hann called the "causes underlying the succession of atmospheric processes."[8] But Hann urged that attention be paid both to the averages and to the deviations from them. This was, Deborah Coen argues, part of a desire to forge a national Austrian identity out of a range of diverse climate zones. To do so required blending distinctive local settings into a harmonious imperial whole. Statistically speaking, this meant attending both to the deviations and to the average, the local and the global. In a real sense, it was the climate averages that made the deviations—and the dynamism they implied—more visible. As a result, and somewhat counterintuitively, Hann's approach ultimately paved the way for a new kind of climatology focusing precisely on the variability within a climate system rather than the stability of individual climatological zones.[9]

When Walker arrived in Simla to take up the post as Director-General of Indian Meteorological Observatories, he walked into precisely this uncertain space between the stability that averaged climate statistics generated and the variability those data could be used to search for. Whether the climate system was seen as inherently stable or inherently variable depended on the perspective of the person inspecting it. Prerequisite to either approach was the concept of a system itself. That system was a product of the Empire as surely as was Gilbert Walker, or the bales of wheat upon which so many depended for sustenance and for profit.

The challenge that faced India and, in particular, the British in India was really a set of challenges. When Walker arrived in 1903, the summer monsoon rains had failed in India for three of the past seven years, in 1896, 1899, and 1902. The brevity, and simplicity, of such a statement belies the grand scale of human suffering it unleashed.

Millions had died. Of the things that were countable in the universe, the number of dead was not among them at that time, so it was impossible to know precisely how many. A *Lancet* article in 1901 put the number who had perished in the past five years at nineteen million, half the total population of the UK at the time and roughly eight percent of that of India.[10] Crops had failed completely across areas totaling more than three times the size of the entire UK.

This was not a natural disaster. The language of "failure" which was (and still is) used to describe the lack of monsoon rainfall in some years implies that the rains were to be expected every year. In fact, variable rainfall—including the complete absence of rain in some years—was a normal, rather than an abnormal, feature of the Indian climate. The monsoons had always come and gone, sometimes providing life-giving moisture, sometimes withholding it. Fam-

FIG. 4.3. An American tourist and an unidentified woman pose with a famine victim, India, 1900. Credit: John D. Whiting Collection/Library of Congress Prints and Photographs.

ines had accompanied such droughts in the past, but the number of deaths from starvation had increased dramatically during British rule. This reached a peak between 1876 and 1878, when some six to ten million had died (across both British and non-British territory).

The British Empire was largely at fault for this harvest of death. In the process of imposing a cash economy on India, the British had dismantled traditional systems of mutual relief and grain storage that had allowed farmers to build up grain reserves during "fat" harvest years that could be drawn upon to make it through the lean years.[11] In the name of productivity, the British had encouraged the destruction of countless individual safety nets. In their place, they had offered ready cash for this year's crop and little else.

In the midst of the famine, Queen Victoria was proclaimed Empress of India. Imperial pomp was blind to the suffering, willfully, even righteously so. Lord Lytton, poet and viceroy of India, proudly wore the mantle of Adam Smith, who had claimed, in relation to the Bengal famine of 1770, that famines were worsened by "improper" and violent government interventions. According to Smith, so-called "humanitarian hysterics" who insisted on sending money for famine relief were in fact contributing dangerously to the possible bankruptcy of India. The best thing to do was nothing. By letting the famine run its "natural" course as quickly as possible, a natural correction in the economic cycle would be effected and, like a series of frequent brushfires that prevent a catastrophic blaze, thereby limit the potential for the worst losses. Famines were in this sense natural events, in social and economic terms, a kind of built-in mechanism to keep the population of India in balance with its size, and, not incidentally, to keep grain prices high. Any attempt to limit the effects of famine, claimed Lytton to the Legislative Council in 1877, only added to the problems of overpopulation.[12] In a brutal bit of human calculus, Lytton pointed out that since the overwhelming majority of those who died in the famines were poor, any policies that had the effect of saving their lives only increased the proportion of the population living in poverty. It might be better that the poor should die, was the unstated conclusion, than left to live lives which were subhuman.

As visible as this failure of government was, British rulers went to great lengths to fail to see what was happening around them. In a state where ten percent of the population had perished from starvation, a glance from the window of the vice-regal train was sufficient to reassure Lord Elgin of the "prosperous appearance of the country even with the small amount of rain that has come lately."[13] Notwithstanding such heroic acts of self-deception, death on such a biblical scale demanded a response. Following the catastrophic loss of life in the so-called Great Famine of 1876–1878, a commission had been established to determine what steps could be taken to avoid such a disaster in the future. Experts in fields such as medicine, economics, and agriculture were consulted, and special regional famine laws, or codes, were written to ensure that aid would be delivered locally in a timely manner. The famine commissioners lamented the inadequacy of meteorology to the task at hand. Whatever clarity the future might bring about the "true periodical fluctuation" in the rainfall, it was painfully clear that contemporary scientific knowledge was sorely lacking as a basis for forecasting. Though famines were inevitable, the depressing truth was that "they will come upon us with very little warning and at very irregular intervals."[14] Forecasts had been issued for monsoons since the 1880s, but the decision was made to cancel them following their failure to predict the absent monsoon in 1901–1902. This decision disappointed many who felt the forecasts were useful even if they were not always accurate. One commentator in the *Times of India* argued that "in a country so essentially agricultural as India the myriad cultivators may not unreasonably ask why such help as the Meteorological Department may be capable of rendering to them . . . is suddenly to be denied."[15] Such requests fell on deaf ears, and instead the commissioners expressed the hope that other technologies of distance that had served the empire would now serve the people. The railroads and telegraph system that were normally used to regulate the flow of commerce—specifically the grain which was the greatest export crop of the Indian empire—would be used, in the case of future droughts, to deliver relief food where it was needed and, by evening out supply and demand, ensure that grain prices did not spike, as they had during the last famine.

In the event, precisely the opposite happened. Trains were used to transport grain not to where it was needed by hungry people, but to where it could be sold for profit. Telegraphic news enabled speculators to corner the grain markets. Local charity was grossly inadequate to the task of caring for large populations of starving people. Desperate parents, unable to feed their children, sold them for pennies each, or, failing that, tried to give them away. One correspondent reported visiting an orphanage where he encountered children whose arms were no bigger than his thumb, and whose ribs showed through their skin "like a wire cage."[16]

Such horrors formed the backdrop against which Walker took up his post on the first day of 1904. As desperate, and even absurd, as it may have seemed to recruit someone like Walker to achieve the seemingly hopeless, there were several reasons to think that Walker was in a better position than anyone ever had been to study and eventually be able to predict the rhythm of the monsoons. He'd been living with high expectations for most of his life. Prognostications of his promise dated back to a mythic mistake he'd made in school, when he'd bungled the declination of a Latin verb. The mathematics teacher who took him in after his subsequent banishment from the classics could not stop marveling at what Walker was able to do, mathematically speaking, with very little effort.

From the age of seventeen, there is evidence that he loved anything that was spinning or turning. A gyroscope he made with his own hands won him a prize in school, and more notice. At Cambridge, he studied applied mathematics with the leaders in the field, J. J. Thomson and G. H. Darwin. In his spare time, he threw a boomerang on the wide green lawns that rolled down to the river from the backs of the great colleges. "Boomerang" Walker could make the curved piece of wood fly far away before it made an improbable, arcing turn and came to roost once more in his hands. It was a noticeable eccentricity at a time when most young men put their bodies to the test in the mosquito-thin rowing boats on the Cam.

Mostly, he devoted himself to mathematics, and in particular to mathematical physics—the study of how objects (those boomerangs) moved through abstract, geometrical space—which formed the backbone of the Cambridge program. In a sense, he pulled it off. At the end of three years of study, he'd not only taken the notoriously difficult Mathematical Tripos exams but come first, living up to the promise that so many—his schoolteachers, his tutors, his coaches, and his parents—had laid a claim to. But the achievement had taken a serious toll on him. He had what was delicately referred to as a "breakdown" in his health, necessitating removal from the location where he'd reached debilitating heights. Three winters at a sanatorium in Switzerland were required to smooth out the mental kinks, the places deep inside him where tension, a necessary quality if one was to marshal numbers at the heights he'd scaled, had become crippling.[17]

John Hopkinson, a fellow Cambridge mathematician turned engineer, had once said that "Mathematics is a very good tool but a very bad master."[18] By the time Walker arrived in India, he'd learned for himself what that meant. Mathematics alone was unwieldy, dangerously consuming, while being simultaneously useless. Boomerangs and their reassuring returns were not enough to salve the hurts that mathematical intensity inflicted on him. In Switzerland, he found succor in ice-skating. The lack of friction, the cold air, and the clear skies rinsed his mind. The arcing boomerang found its echo in the curve of his skates on the ice. Inside his mind, some reciprocal curve began to grow, rebuilding the parts of him that had been broken by too much study, too much mental tension. He skated his way through, and eventually out of, his breakdown. He spent several years back in Cambridge as a college lecturer, seeking suitable material with which to ballast his flightier tendencies, something weighty enough to keep him from flying off into a mathematical abyss. Electrodynamics, and a problem suggested to him by a senior mathematician, kept him tethered for a while.

That time ended when, aged just thirty-five, he was recruited to be the new head of meteorological observatories in India and to join the cadre of scientific professionals who populated a thin but growing strata in the great laminated system that was the British Raj. Those who had come before him in the meteorological field had tried, in their way, to master the weather from the ground up, with maps of storm systems and theories about the effect of snowfall in the Himalayas on next year's monsoon. But the conclusion to which they'd come was that the connections between aspects of the weather that mattered most to India were too complex to reveal themselves to even the most intuitive and insightful of scientists. In the past, the basis of meteorology had always been physical. Scientists had always tried to picture things visually, to imagine the way different masses of air might interact to push and pull each other around the ocean of air that was the atmosphere (to say nothing of the masses of moving energy in the ocean itself). They had failed, and now they hoped that someone like Walker, for whom numbers acted like a lever with which to pry open otherwise closed systems, could do the same for India. On hearing news of Walker's appointment, Cleveland Abbe wrote to congratulate him and expressed his hope that "by suggesting a new class of problems, your thoughts may be centred on dynamic meteorology, to the great advantage of this difficult branch of science."[19]

As it happened, opening up India to Walker's penetrating gaze was no different than offering him the world.

By 1904, both the British Empire and the discipline of meteorology were edging toward the farthest boundaries of the planet—they were nearly, if not yet completely, global in extent. The British Empire was close to its peak of influence and power, when it encompassed nearly a quarter of the earth's landmass and a fifth of her population. In India, its largest colony by far, the British controlled an area of some 1.5 million square miles, ten times the size of Britain itself. This kind of lopsided rule was inherently precarious, as the bloody Mutiny of 1857 had shown.

The challenges the Empire and meteorology faced were remarkably similar. Both sought to understand and control a set of unruly phenomena unfolding in locations that were often remote from the offices where calculation and coordination occurred. The notion of imperial meteorology, championed by *Nature* editor Norman Lockyer, was, therefore, something of a redundancy. Empire *was* meteorology, and meteorology *was* empire. To put it more practically, as India's finance minister Guy Fleetwood Wilson memorably did in 1909, the "budget of India is a gamble in rain."

Only with the leveraging power of certain technologies was British rule in India even thinkable. Much has been made of the importance of railways, telegraphs, and steamships in drawing the Empire together across time and space. Just as essential but often overlooked were the tools of bureaucracy itself. These took the form of central offices where information could be gathered, sorted, and acted upon. Such offices were the nodes of the great imperial network. They reached their apotheosis in London, but were necessarily to be found also in Calcutta, in Simla, and in remote field stations from which telegraphic messages were sent and received. In these small and well-organized spaces, a few workers with the ability to move information around with as little friction as possible could contribute to the governing of millions of subjects of the crown.

Thanks to the power of technology and bureaucracy, distance, once a foe to be vanquished, became something rather more interesting and much more valuable to the British Empire. Rather than a challenge to the exercise of power, it came to be seen as a mark of that very power. A tiny post office on an Indian tea plantation, set beside a stream in which an elephant might peacefully bathe, could reveal itself to be, on closer inspection and by dint of a small wire emerging from it, a node in a global network of imperial connection and control. As a result of scenes such as this, often reproduced in imperial gazettes and albums, distance became the leitmotif of the Empire, an expanse on which the sun famously was given no opportunity to set.

Distance wasn't just symbolic of the great power of the Empire, it also created value where none had existed before. Reliably fast steamships that muscled their way across the seas meant that English citizens could eat bread made from Indian grain grown year-round (or nearly so) at a comfortable and safe distance from the land in which it was grown, from both the sunlight and the rains upon which it depended. India became both Britain's bread box and its money box. By 1904, it was Britain's greatest source of imported goods and the largest market for Britain's own exported goods.[20] India's value to the Empire arose not in spite of, but because of, its distance from London.

Distance was what made the Empire work. It was as much a part of the logic of its success as any local control. In many ways, it was inevitable that Walker's greatest achievement would be the discovery of something he called *world weather*. The greatest distances the world could offer were available to Walker, and he took them and put all his skills to bear in meeting the correspondingly enormous challenge that had brought him from the peace of Cambridge to the monsoon wars.

The very meteorological facts that made India such a challenging place to govern made it singularly ripe for meteorological investment and study. India offered a fantasy geography for the meteorologist. This was partly a matter of scale. Everything in India was oversized. Conceptually separated from its neighbors in the region by dint of its special relationship to Britain, it was also physically separated by the extreme vertical boundaries of the Himalayas at its northern edge, by the coasts on its east and west sides, and by a southern tip that, reaching to the equator, tapered into nothingness. Straddling a quarter of the earth's latitude, India demonstrated an enviable range of climatological phenomena. Its scale and climatological features meant that unlike Britain, where weather varied from day to day, in India the weather conspired to generate longer-term patterns: months, rather than days, formed convenient units of measurement. This made calculation vastly more manageable. As a result, India was a place where the patterns of the weather could make themselves more legible than almost anywhere else on the planet. Walker's predecessor Henry Blanford wrote without irony of India that "Order and regularity are as prominent characteristics of our atmospheric phenomena, as are caprice and uncertainty those of their European counterparts."[21] This was partly a matter of geographical extent. India was a place where "general laws have a sufficient space to produce general results," and so-called "disturbing influences are regular and well-ascertained."[22] For anyone who had ever grown first irritated and then maddened by the changeability of English weather, India presented a compellingly bold picture. Inundations were almost normal in parts of the country, while desert conditions prevailed elsewhere. More than 460 inches of rain fell in an average year on the village of Cherrapunji in the Assam hills, while parts of the Upper Sind garnered less than three inches. Impossible things seemed to happen frequently in India. In the wettest regions, it wasn't uncommon for twenty-five inches of rain to fall in a single day—the same amount that normally fell on London in a year.[23] During extremely hot weather, instruments recorded negative readings for humidity in some places. It was common for cyclones to hit the Indian coasts that were stronger than any which had ever been experienced in Europe.

This meteorological profusion took many and complex forms, but the most dominant was that of the monsoon, an alternating pattern of dry land winds that persisted for half the year and high humidity, cloud, and heavy rainfall that lasted for the other half. During the cold season, from October to April, the winds blew in dry and cold from the northeast. They reversed direction from May, bringing from over the oceans the wetter air that bore the heavy rains that fell from June to September or October.

The monsoon was a perfect example of the paradoxes of India. The very thing that caused so much suffering—the unpredictability of the monsoons—might turn out to be the key that could unlock the secrets of the weather more generally. The monsoon was as strong a signal as a meteorologist could ask for, writ as it was in the suffering or prosperity of millions, dutifully reported by those thousands of rain gauges and the busy barometers and thermometers whose readings were also faithfully recorded. And strong signals were the best chance that meteorology had of transforming itself into a more reassuringly predictive science. Norman Lockyer made it seem almost self-evident: "surely in meteorology, as in astronomy," he urged his fellow scientists, "the thing to hunt down is a cycle." Geography should be no obstacle, and indeed need not be, given the great reach of the British Empire. If a cycle is "not to be found in the temperate zone, then go to the frigid zones, or the torrid zones, and look for it," urged Lockyer, "and if found, then above all things, and in whatever manner, lay hold of, study it, record it, and see what it means."[24]

If the monsoon's variable cycle was about as hard to miss as an elephant at close range, it was a more difficult matter to determine what caused the rains to come when they did. The starting point was the sun, the only thing more visible than the Indian monsoons. It had already offered up a well-characterized cycle of its own, during which its spots waxed and waned. These black spots, first noticed by Galileo, had been studied since by scientists seeking to understand what effect they might have on the earth. In the eighteenth century, astronomer William Herschel compared the historical index of grain prices in Adam Smith's *Wealth of Nations* with sunspot data, looking for correlations. In the 1830s, the Magnetic Crusade to map the magnetic currents of the earth had sent observers with magnetic instruments to the four corners of the globe. They'd hit the jackpot when they discovered that the earth's magnetic field fluctuated in time with the sun's. Interest in sunspots took off further in 1850, when Heinrich Schwabe published nearly twenty-five years' worth of daily records he'd made of sunspots —the best data set so far—which he used to identify a ten-year cycle of waxing and waning spots. That figure was soon revised to eleven years, and the sunspot cycle seemed even more likely to have definitive impacts on Earth. More grist was added to the mill in 1859, when a very strong solar flare had caused magnetic instruments to go haywire, sent telegraph communications offline (and even set some telegraph stations on fire), and generated visible aurora even by the equator. Thanks to the sense of urgency occasioned by such events, funds were made available to build a series of special observatories that were meant specifically to observe the sun and to collect and analyze data about possibly related phenomena on earth (on Piazzi Smyth's expedition to Tenerife, he carried many requests from leading scientists to make solar observations). Caught up in the sense that certain natural mysteries were on the cusp of being revealed, physicists sought, and to a certain extent found, links between sunspots and magnetism, sunspots and temperature, sunspots and wind, and sunspots and rainfall. That these relationships could often be summed up in almost disarmingly simple terms was part of their allure. The links seemed obvious. Charles Meldrum, the government astronomer at the official observatory in Mauritius, summarized his own findings thus: "many sunspots, many hurricanes; few sunspots, few hurricanes."[25]

But despite the energy that went into solar physics, by the early twentieth century no further direct physical connections had been discovered between the earth and the sun to rival the findings of the Magnetic Crusade. Interest gradually waned. Sunspottery, as detractors called it, came to look dangerously like a dark art, its practitioners finding patterns where none rightly existed, in a morass of confusing detail.

Among scientists, a small group remained faithful to the search for links between the sun and the earth. These cosmic physicists were less interested in cracking a secret code of nature than they were with understanding the fundamental physical connections between phenomena. Unlike most physicists, who concerned themselves with the behavior of electricity, magnetism, and heat at very small scales, these physicists probed nature at the very largest of scales, that of the solar system and beyond, on the assumption that "a force not less universal than gravity itself, but with whose mode of action we are as yet unacquainted, pervades the universe, and forms, it might be said, an intangible bond of sympathy between its parts."[26] It was undeniable that the sun affected some aspects of earthly phenomena. The hunt was on to figure out what precisely was the nature of the force "not less universal than gravity" which was responsible for such effects. They were convinced that physical connections between the earth and the sun (among other celestial objects) were profoundly important to the unfolding of meteorological, magnetic, and electrical phenomena on earth. Though the distances they worked with were great, they thought in surprisingly sensuous terms. Like lovers, the sun and the earth were exquisitely attuned to each other. "Mutual relations of a mathematical nature we were aware of before," wrote two leading cosmic physicists, "but the connexion seems to be much more intimate than this— they feel, they throb together, they are pervaded by a principle of delicacy even as we are ourselves."[27] Tiny, ramifying perturbations could arise anywhere in the solar system, not just on the sun itself. Like the trigger of a gun, small changes in the gravitational fields of other planets in the solar system could cause sunspots that could themselves have huge effects on earthly weather. The sun was therefore able to produce incredible variation in earthly weather "by falling at different times on different points of the aerial and aqueous envelopes of our planet, thereby producing ocean and air currents, while, by acting upon the various forms of water which exist in those envelopes, it is the fruitful parent of rain, and cloud, and mist."[28] Such a passionate belief in these connections gave cosmic physicists patience and hope in spite of the lack of results to date. The seemingly impenetrable variability of the weather was a measure "not of its freedom from law," wrote Lockyer and Hunter, "but of our ignorance."[29] All natural things, including that most fickle of phenomena, rainfall, would eventually be shown to obey the laws of nature. Only more time was needed.

And data, lots and lots of data. Of that, at least, Walker had as much as, if not more than, he could have hoped for. Walker was not a cosmic physicist, by training or inclination, nor was he (anymore) susceptible to the fever for cycle-hunting. He was, instead, a man for whom numbers were tools that could be put to particular uses. The disciplining of numbers was essential. Just as Walker had put himself to the ultimate test of disciplined study as a student at Cambridge, so he submitted his numbers to the test of reliability and meaningfulness.

He'd inherited the concerns of those who'd come before him, and he would be conscientious in addressing those concerns. It would be wrong to say he wasn't guided by the past. In many ways, the questions he asked of his numbers were questions others had already raised, about the relationship between distant phenomena, about the way in which things that are very far apart can, indeed, be linked. This characteristically imperial notion was both enabled and promoted by all the structures—the railways and telegraphs, and bureaucratic structures—which made the empire possible. Walker was, like anyone, influenced by the world around him, by what had come before, and by what he was hired to do in the moment.

First, he needed to gather the numbers themselves. That in itself
was not difficult. He was in charge of the most advanced meteorological network in the world. Numbers—corresponding to facts about
the weather—streamed into his office day after day, month after
month. No data monsoon or failure thereof afflicted the Director-
General of Observatories. In 1907, for example, his office in Simla
received the records of rainfall from 2,677 rain gauges across India.
He also received readings from several dozen meteorological observatories of pressure, temperature, and wind speed taken at eight-
hourly intervals and, in some locations, recorded continuously by
automatic instruments. He knew he needed data from the oceans if
he was to crack the monsoons, and so he sent two full-time clerks
to Calcutta and Bombay whose sole job was to visit ships as they
came into harbor for the purpose of copying their meteorological
logbooks and calibrating their barometers. The atmosphere—that
ocean of air—wasn't as easy to access, but it was critical to create a
three-dimensional map, if possible, of the air currents that brought
rain or dryness. By 1904, when Walker took up his job, it was
generally agreed that more information on the middle and upper
atmosphere was urgently needed, and should be acquired by any
means—kites, balloons—necessary.[30] Walker sent balloons and kites
up from Belgium, and over the Bay of Bengal and the Arabian Sea.
They flew as high as 2.5 miles into the atmosphere. From Simla, he
sent up gutta-percha observation balloons that carried ultra-light
instruments. These had to be recovered for their data to be useful,
and he attached cards to the balloons, promising a reward for their
safe return. The previous Director-General of Observatories, Henry
Blanford, had pointed to snowfall in the Himalayas as an important
factor in the monsoons, so Walker arranged for large-scale photographs to be taken of the snowfall visible from Simla, which could
be compared year on year.[31]

And he corresponded. He set up telegraphic and postal corre-
spondences with fellow observers around the world. The office
in Simla, with its characteristic telegraph wire leading out of its
windows, received weekly telegraphs keeping Walker informed of
weather conditions at the Royal Alfred Observatory in Mauritius, a

key location from which monsoon winds blew. Departmental observatories in Zanzibar and the Seychelles provided much-needed data on the Indian Ocean. For the southwest monsoon, he corresponded with Zomba, Entebbe, Dar es Salaam, Cairo, and Durban in Africa; with Perth, Adelaide, and Sydney in Australia; and with Buenos Aires and Santiago in South America.

All this data was promising, and seemed necessary. It was also potentially fatal to the dream of solving the monsoon mystery. Too much data could easily prove incapacitating. This was the dilemma. To understand the phenomenon, it needed to be observed. But it wasn't clear precisely what the boundaries of the monsoon were. Where it started and where it ended was part of the answer Walker was seeking. So he needed, as had others before him, to cast his net wide. But the wider he cast, the more numbers he caught, the harder it would be to find the elusive signal amid all that noise.

"There are undoubtedly too many observations," noted John Eliot, "and too little serious discussion of observations." Instead of accumulating observations without consideration of how they might be used, the time had come for investigation of causes to "direct and suggest the task of observation." A natural feature of a more thoughtful observing regime would be to consider allied sciences alongside that of meteorology—"there are undoubtedly definite relations between certain classes of solar phenomena and phenomena of terrestrial magnetism" and who knew what other links might be uncovered.[32] What Eliot suggested was the creation of a central organization where observations taken throughout the British Empire could be compared.

As Arnold Schuster, a prominent cosmic physicist, put it, "observations are essential, but though you may never be able to observe enough, I think you can observe too much . . . It would not be a great exaggeration to say that meteorology has advanced in spite of the observations and not because of them." There was always the danger that data collection would become an end in itself and science would become nothing more than "a museum for the storage of disconnected facts and the amusement of the collecting enthusiast."[33]

Before the matter of the proper place of observation in meteorology could be settled, the nature of meteorology itself needed to be resolved. What meteorology was, exactly, was up for grabs. Should it consist of prediction? Of observation? Of theorizing? Or, as seems reasonable, a mix of the three? But if a mix of the three, in what sort of hierarchy should these different approaches to the atmosphere be ordered? The question of whether prediction could precede theorizing was a potentially explosive one, as the cancellation of weather forecasts in response to FitzRoy's death makes clear. Some felt strongly that prediction without proper theory was a dangerous undertaking—for both the public who might be provided with faulty forecasts and, just as concerning, for the scientists who were wary of exposing what they saw as the weakness (or immaturity) of their discipline. American meteorologist Cleveland Abbe spoke for this point of view when he wrote in 1890 that "hitherto, the professional meteorologist has too frequently been only an observer, a statistician, an empiricist—rather than a mechanician, mathematician, and physicist."[34] Others felt just as strongly that theories based on too few observations were as useless as observations unleavened (as one commentator put it) by theory. As noticeable as the differences between what might be called theory- versus data-led meteorology might have seemed, there was not as much separation between these attitudes as might be thought. Indeed, depending on the problem at hand, the same person might advocate first a data-driven, and then a theoretical approach. Julius Hann, for example, did more than anyone to establish the descriptive, empirical tradition of climatology, but he also applied thermodynamics—a highly theoretical field—to problems of atmospheric phenomena.[35]

Just as Hann saw climatology as a helpmeet for meteorology, so others argued that physics was a needed stiffener that could transform the science of the atmosphere into a true science. The scales to which these disciplines directed their energies were largely as distinct as their methods. While climatologists set out to encompass the globe with imperial maps keyed to resource production and extraction, meteorologists focused instead on devising physical theories that could be regional, local, or even, in the case of clouds, hyperlocal in nature.

As these differences between meteorology, climatology, and an incipient physical geology such as informed the ice age debates indicate, the concept of change was itself unstable from the middle decades of the nineteenth century through to its close. Which kind of changes could be looked for, by whom, and using which tools had become very public and very controversial questions by the end of the nineteenth century. What it meant to be a science— how much it could be a function of data gathering and how much it required theory—was a primary question from which everything else, including what even counted as data, emerged.

These disciplinary anxieties formed the backdrop to Walker's conundrum. How was he to escape the confines of Schuster's meteorological "museum," full of musty and disconnected facts? What Walker realized, thanks in part to the work of those who had come before him, was that solving the mystery of the monsoons would require two things. First, he would need to shift the scale of inquiry from local, regional, or even pan-regional studies to truly global surveys of world weather. Second, and just as importantly, Walker realized that he needed to ditch cycle-hunting in favor of something with a qualitatively different mathematical basis. He saw himself not as a hunter in search of one charismatic meteorological megafauna—the singular "link" between one cycle and another— but as a surveyor charting the landscape of the weather itself.

Here is where Walker's ignorance of the weather may have been his greatest asset. Without any pre-existing assumptions about which aspects of the atmosphere might have the most bearing on the monsoons to guide him, he realized he needed a tool with which to evaluate all factors and to help him determine which, if any, were the most important. The tool he had was statistics. Specifically, he developed a technique for calculating what he called the reliability of the correlation coefficient between two factors. What this meant was that he had a device by which to sift the vast mountain of data. Before Walker's innovation, the best tool cycle-hunters had was visual. They plotted charts comparing one signal against another (such as barometric pressure against sunspot appearances) and looked at the resulting curves to see if any pattern emerged—either an especially close fit or an especially poor fit, evidence perhaps of an inverse relation. Walker realized he could sharpen a device developed by statistician Karl Pearson, called a correlation coefficient, to sift through numbers statistically. Pearson's correlation coefficient was a tool for identifying patterns—degrees of correlation—that linked two sets of data. This was a very helpful tool for sorting the vast reams of statistics which came flooding into Walker's office.

A problem arose when it came to looking for real patterns in weather data: Pearson's correlation coefficient tool was sometimes too good at finding patterns. When comparing two sets of random data, there is always a certain likelihood that you will find a relationship between them. The same is true when comparing real data, such as, for example, barometric pressure in different parts of the world. Pearson's tool was unable to distinguish between the real correlations—that is, those that indicated underlying physical connections—and those that arise purely as a function of the quantity of data being compared. When comparing dozens, or even hundreds of data sets, as Walker was, the chance of false positives is large. Walker's tool provided a measure of just how much correlation was required to balance out the likelihood of false positives in large data sets.

By applying his criterion of reliability to Pearson's correlation coefficient, Walker was able to generate a quantitative measure of the likelihood that a correlation between two series of numbers was not due to chance. Instead of eyeballing a series of curves, Walker could rank relationships numerically and identify which were statistically robust and therefore likely to reflect something happening in the real world and those that were less strong and therefore more likely to be merely random. His technique was both more accurate and vastly more efficient at sorting through the huge data sets with which he was confronted than that of his predecessors. It was, as Napier Shaw, another leading researcher, recognized, a "kind of searchlight for sweeping the meteorological horizon from some selected point. The principal features of the otherwise invisible landscape, which in this case extends over the whole globe, can thus be located."[36]

The landscape of Walker's investigation was global. This was, again, as much a function of Walker's ignorance as it was a calculated decision. Without a clear sense of where to shine his spotlight, he needed to shine it everywhere. Any correlation he found was, as Napier Shaw put it, a "very sensitive plant, it is much easier to kill one than to make one; whatever happens in the way of accidental errors, it must suffer."[37] That was the idea. If Walker were to find real relationships in the vast sea of data, he had to be merciless with any putative links. Only the strongest, most statistically resilient could be allowed to survive. These could lead the way for scientists who had a physical theory of the circulation of air, wind, and rain to explain what Walker had merely indicated.

And what he had helped reveal was this: There was something called world weather. It consisted in large regions of alternating high and low pressure that spanned the globe and changed with the seasons. There had been theories before of what had been called the general circulation of the atmosphere, dating back to Hadley's theory of the trade winds in the eighteenth century. More recently, a spate of work done during the 1880s and 1890s, drawing on the same sorts of telegraphic correspondence networks that Walker did, had begun to pick out a series of such oscillatory, or seesaw, relationships between areas of characteristically high or low pressure. These papers, many of them by cosmic physicists, blended the tools and approaches of the cycle-hunters with those of the physicist accustomed to thinking about physical connections between matter. They generated maps, often of pressure but also of temperature, which demonstrated intriguing, even astonishing, connections between distant parts of the earth's atmosphere. The term *oscillation* was used early on to describe the inverse relationship between pressure in different parts of the globe that many of these studies found. Léon Teisserenc de Bort, an architect of universal cloud studies, had shown that there was a relationship between the average pressure in Europe and that in certain "centres d'action" in Iceland, the Azores, and Siberia. Henry Blanford had done similar work for the Southern Hemisphere, showing that pressure in India, Siberia, and Mauritius was linked. H. H. Hildebrandsson, a round-faced Swede, had gone much further with his monumental series of five memoirs present-ing a ten-year run of average monthly pressure data from no fewer than sixty-eight locations from all quarters of the globe. He used this data to push even further from these still hemispheric centers of action to suggest that there were what he called "intimate relations" between *all* of the centers of action on the globe.[38] And finally, Hildebrandsson and de Bort's Cloud Atlas of 1896 had shown that it was possible to leave behind what Julius von Hann called "church steeple politics" in meteorology (what could be seen from a church tower) and move toward ambitious, global projects.[39] Clouds, to state the obvious, obeyed borders not at all, so any project to map them had to be similarly wide-ranging.

Here, then, was the landscape upon which Walker could shine his searchlight. From a tradition that stretched centuries into the past, which had built up interest in and knowledge of storms, to more recent attempts to collect and compare data at hemispheric scales, Walker had managed to arrive at precisely the right moment to sub-mit a truly global data set to scrutiny. Like Blanford, Teisserenc de Bort, and Hildebrandsson, Walker found evidence of oscillations in the pressure data he'd collected. But while they had been limited by their visual techniques to making vague statements about the nature and degree of these connections, Walker's correlation coefficients allowed him to eliminate those connections that were less mean-ingful. He found 400 significant relationships—correlation coeffi-cients worth paying attention to.[40] Subtracting the spurious connec-tions left him with "three big swayings," or inverted relationships between pressure. The biggest was between the Pacific and Indian Oceans. This Walker named the Southern Oscillation. Two smaller swayings, between Iceland and the Azores and between parts of the North Pacific, he named the North Atlantic Oscillation and North Pacific Oscillation.[41] In these locations, pressure existed in inverse relations. When the barometric pressure rose in Iceland, it seemed to fall in the Azores, and vice versa.

One of the first questions to which he put his correlation coefficients was that of sunspots. In a 1923 paper, he demonstrated that there were no meaningful correlations between the eleven-year sunspot cycle and that of the monsoons.[42] He seemed to recognize the discomfort, and even disappointment, he may have caused. It was natural, he acknowledged, "after long ages of belief in the control of our affairs by the heavenly bodies," to believe in natural cycles. But the urgent need for good monsoon forecasts, and the terrible suffering the famines had caused him, had nevertheless driven him to "replace instinct by valid quantitative criteria."[43] Eliot's gamble in hiring Walker had paid off. Sort of. For even as Walker brought the edges of meteorology and empire to their ultimate endpoint—the entire earth—his achievement also represented a retrenchment and scaling back of ambitions. In gaining world weather, he had sacrificed the cosmos. Taking away the hope that secret cycles might unlock the monsoon was just about acceptable if, in return, Walker could offer something better.

That something better was, of course, his original goal of predicting the monsoons. The monsoon forecasts, which had begun in the 1880s and had been suspended in 1902 following the disastrous famines, had been reinstated on the basis of Walker's findings. Walker's predecessor, Eliot, had emphasized how dangerous "the striving after perfection in short-period forecasts was."[44] There was too much imperfect information and experience with failure to treat forecasts as anything other than probabilities. But Eliot's cautious words were hard to hear against the background of famine and economic imperatives, and the government pressured Walker to publish forecasts once again. Walker was the first to be cautious, and even critical, of the forecasts, which he emphasized were only as good as the correlation coefficients he was able to find. These varied from year to year, sometimes dramatically. He urged that forecasts be issued only with strong provisos. Foreshadows, rather than forecasts, would, he thought, be a more appropriate, modest name for them.[45] But the stronger term had stuck, and the tendency, or desire, for these pronouncements to be powerfully predictive was as great as it had ever been. There were some successes in prediction, but it seemed there were just as many failures, and it was embarrassing, after so much time and expense, and in the face of such evident need, that professional meteorology was often unable to offer better prognostications. The fear of making inaccurate predictions could lead to an absurd state of affairs where experts were less adept at forecasting than simple folk. It was unfortunate, commented one writer, Charles Daubeny, when "the untutored peasant sometimes would seem to possess an intuitive insight, whilst the philosopher, although he may plume himself on his acquaintance with the general laws of atmospheric phenomena, is often at a loss to unravel the entangled skein of effects connected with it which daily observation brings before him." Meteorologists were damned if they did and damned if they didn't. A bad prediction could tarnish their name, while too much reserve was also unacceptable. Daubeny continued that while charlatans had no compunction about making predictions, "a Herschel or an Arago declare themselves incompetent to anticipate what may chance to supervene within the space of the next four-and-twenty hours."[46]

The ironic fact of the matter was that it was easier to use the monsoon to predict what would happen elsewhere in the world than it was to predict the rains themselves.[47] Why that was the case, Walker the mathematician was unable to say. Luckily, though the monsoon continued to flummox and tantalize hundreds of millions of Indian farmers and those who relied upon their grain, no more famines occurred on the scale of the terrible death and suffering that had preceded Walker's arrival. Changes in economic and social policy on the part of the British, and a run of good monsoon years, were to thank for that.

If Walker had failed in his primary objective of predicting the monsoon through statistical means, he had also failed to provide any physical explanation for the discovery he'd made. It was, in a way, like throwing a boomerang without understanding the physics beneath it. In that case, his lack of knowledge hadn't kept him from excelling, but he'd nevertheless been driven to try to describe precisely how the device worked. Though he'd embraced his task in India using the most effective means that were available to him, he never forgot what he'd lost in the bargain. In a 1918 lecture to the Fifth Indian Science Congress, he emphasized how important it was to have a grasp of the fundamental principles driving phenomena under study. "What is wanted in life," he urged the students, "is ability to apply principles to the actual causes that arise . . . When Pasteur as a chemist was asked to find a remedy for the pest that was ruining the French silk industry, he knew absolutely nothing of silkworms; yet he solved the problem, and it was general understanding of Nature's methods that brought him success."[48] Walker knew better than anyone that that general physical understanding was precisely what was missing from the world weather he had discovered.

Just as Walker had failed to find a way to predict the monsoon, the larger project of melding meteorology with astronomy that went by the term *cosmical physics* had, by the end of World War I, largely faded from view. In its place was a new branch of meteorology. Instead of trying to link the heavens and the earth, as cosmic physicists had, this new type of meteorologist tried to link the lower atmosphere, to which most meteorological measurements had long been confined, with the upper, which was becoming gradually more accessible. Charles Piazzi Smyth's expedition to Tenerife was an early example of the push to establish mountaintop observatories that, in addition to offering better views of the stars, made it possible to take readings of the upper air. Mountains had obvious drawbacks when it came to tracking the movements of a free-flowing atmosphere. After a series of spectacular and dangerous balloon ascents into the upper atmosphere, notably by English meteorologist James Glaisher, researchers sought safer ways of taking readings of the air high overhead. One way was to observe the motions of clouds, as the organizers of the International Cloud Atlas understood. But these observations could only reveal so much about the atmosphere. More precise data would require sending instruments themselves into the skies. Kites and unmanned balloons soon became the prime instruments for plumbing the ocean of air. In the late 1890s, Teisserenc de Bort, who had retired from his post as director of the Central Meteorological Office of France, established a meteorological field station at Trappes, southwest of Paris. There, he pioneered techniques for launching the large and delicate balloons needed to reach the upper atmosphere, using a large hangar set on a rotating platform, which could protect the balloon from ground winds until it was safely launched. Using this apparatus, and a self-registering device to record temperature, pressure, and moisture, Teisserenc de Bort carried out dozens of soundings in the years around 1900. The traces recovered from the self-registering device—which scratched its readings into lampblack that was impervious to the damp conditions—revealed a new aspect to the atmosphere. The temperature of the atmosphere fell in a uniform manner until the balloon reached some eight kilometers high, at which point it stopped decreasing. In 1902, Teisserenc de Bort named this region of the upper atmosphere the stratosphere, and coined a new phrase for the layer closest to the earth, the troposphere.[49]

Walker himself was well aware of the need to understand the upper atmosphere better. "I think the relationships of world weather are so complex that our only chance of explaining them is to accumulate the facts empirically," he wrote at the end of his life, "and there is a strong presumption that when we have data of the pressure and temperature at 10 and 20 km, we shall find a number of new relations that are of vital importance."[50] During his tenure as Director-General, he established an upper-air observatory in the northern plains of India, at Agra. Starting in 1914, a ten-year experimental program was carried out. Among other things, the balloons sent up by Walker and his men showed that the stratosphere—the zone of constant temperature—started much higher in the atmosphere above India than it did in Europe.[51]

Walker left India in 1924 after twenty years of service. His achievements (including helping hire increasing numbers of Indians into the Met Office) were lauded, he was awarded a knighthood, and he took up a position as professor of meteorology at Imperial College. He soon joined the Imperial College Gliding Club. And though he complained that his reflexes were not sharp enough for successful gliding, he accompanied the younger gliders on several expeditions in the South Downs. He sometimes took his boomerang with him and sent the device flying far above him in the gentle air of southern England before it began its perfect return, vibrating cleanly through invisible turbulence before coming to rest in his long, elegant fingers.

He never did learn what caused the monsoons. In 1941, nearly twenty years after he had left India, he received a letter from then-director of observatories Charles Normand, informing him that the monsoon forecast for that year based on Walker's work was "little or no better than will be given by the intelligent layman who knows no meteorology but does know the monsoon frequency curve." Normand was, understandably, reluctant to issue an official forecast on this basis. "I much prefer not to speak," he explained, "unless the correlation forecast is appreciably more useful than the intelligent layman's." Walker could only concur. He'd never placed much store in the forecasts himself. "I fully agree with your policies of not making fuss about monsoon forecast," he wrote back.[52] The truth, as Walker was the first to admit, was that the Southern Oscillation was "an active, and not a passive feature in world weather, more efficient as a broadcasting than an event to be forecast," as Normand put it.[53] By 1950, the dream of forecasting the monsoon had been, if not fully abandoned, put on indefinite hold. Not only was it clear more data was needed, it was looking increasingly possible that data alone would never be enough. S. K. Banerji, who became the first Indian head of the Indian Meteorological Department in 1945, was clear-eyed about the limitations of an effort into which an "enormous amount of labour" had been poured. "The results obtained are not satisfactory. We do not, however, know yet all the factors which control the Indian rainfall. . . . It seems unlikely that a complete solution will be achieved in the near future. It is possible that part of the seasonal rainfall is not predictable in advance."[54]

Walker never presumed that success was inevitable. Still, it is hard to read this story without feeling a sense of disappointment. The man who had definitively ended the search for correlations between sunspots and monsoons had been unable to find his own holy grail—a means of predicting the monsoon. In the process of looking, he had discovered something very important, a way to begin investigating the links between distant parts of the global atmosphere by validating which statistical connections were most likely to indicate physical connections. The exact nature of those physical connections remained unclear to Walker, and indeed, impossible to discover by the methods he had developed. It was only in 1969, ten years after Walker's death, that a further veil on the mystery of the monsoons was removed when the Scandinavian meteorologist Jacob Bjerknes showed what was missing from the landscape of Walker's world weather.[55] The ocean was the enormous elephant in the room. It provided the necessary other half of what Bjerknes showed was a grand global cycle whereby ocean temperatures affected the temperature of the air above it. He named this cycle of east- and westward motion the Walker Circulation. Its basic mechanism was this: Cold water that welled up from the depths of the eastern Pacific cooled the air above it, preventing it from rising and therefore allowing it to be blown westward by the trade winds, where it eventually warmed enough that it rose above the western Pacific. It was then able to return eastward in the upper atmosphere, where it closed the circle by sinking back down over the Pacific. Variation in the degree of cold water that upswelled, unexplained in 1969 (and still mysterious today), seemed to be the reason why some years the circulation "failed" to bring monsoon rainfall to India.

Walker and Bjerknes did work which, laid end to end, would eventually solve at least some of the mystery of the monsoon. This tells us something important about the way our understanding of the earth has evolved. It is the movement between observation, calculation, and theorizing that produces insights. No prescription can set the order in which these different ways of knowing can, or should, proceed. And no reliable method has yet been devised which can forecast from which quarter an important new piece of work will emerge. Too late for Walker, the great arcing trajectory of his monsoon research did, eventually, find its return. The monsoons were part of a global system by which heat travels through the oceans and the atmosphere, making its way around the complexities of water and air as surely as the boomerangs Walker threw found their way back to him.

5

HOT TOWERS

Joanne Gerould, aged twenty-one, stood in front of a roomful of aviation cadets at the University of Chicago. The year was 1943 and the United States was at war. Though she was young, and more unusually still, a woman, she had good reason to be standing where she did. She knew more than the cadets did about the movement of air and moisture through the atmosphere. That one thing was reason enough to grant her the authority to lecture them. The men needed to learn, as quickly as possible, the fundamentals of weather forecasting.

What the young woman needed was less clear, but she already had a strong sense of what she didn't want. She didn't want to be dependent on a man. She'd learned from her mother the emotional damage it could do to be bright and unable to follow one's dreams. Her own mother had trained to be a journalist but never managed to pick up the threads of that ambition after she'd given birth to Gerould. She'd taken out her frustrations on her daughter, and Gerould, in turn, had struggled with the weight of that bitterness. She'd searched for escape, finding it in the teeming confusion of the marshy estuaries in which she'd played as a child on Cape Cod, in the coastal waters upon which she'd sailed and in which she'd swum, and finally, in the skies through which she'd flown. Aged just sixteen, Gerould had obtained a pilot's permit. It was both a metaphorical and a very real form of escape, out and up, into the skies.

When it came time to choose a college, Gerould had followed the same urge, flying away from her home in Cambridge, Massachusetts, and from Radcliffe, where both her mother and grandmother had studied. She went west, to the University of Chicago. There, a course that included plenty of science classes appealed to her. She thought she might study astronomy. But the time was right not for the study of the heavens, but of earthly skies. World War II was, famously, an airman's war. Flat navigational charts, plotted with straight rulers, were set aside in favor of spherical globes on which bits of string were laid to trace the curving paths taken by planes that found their target with unerring directness. Old European battle lines looked set to disappear in the face of the new Pacific arena and the northern pathways that touched the whitest parts of the planet. The geography of the world looked as if it could be remade by these airmen, and the countries for which they flew, if only they could win the war.

At the start of the war, the Germans had more than 2,700 trained meteorologists available to advise their pilots on how to stay safe in the air. The United States had just thirty.[1] To rectify this alarming imbalance, the air force had gone straight to the person who could make the fastest and greatest difference. Carl-Gustaf Rossby was a maelstrom of meteorological theory and administration, a thinker and a doer, with energy to spare. A Swede, he'd received his own meteorological education in Bergen, Norway, at the time and place in which meteorology had come of age professionally and started to deliver on promises long in the making.

In Bergen, a man named Vilhelm Bjerknes had managed to cleave the practical necessity for daily forecasts to the clarifying mathematical equations of physics. The theories of the weather that Bjerknes had helped develop, and which Rossby had learned better than anyone else, were well suited to explaining the skies above Scandinavia. Having come of age during World War I, these men naturally saw the battlegrounds of northern Europe projected on the cold skies overhead. "We have before us," wrote Bjerknes, "a struggle between a warm and a cold air current. The warm is victorious to the east of the centre . . . The cold air, which is pressed hard, escapes to the

FIG. 5.1. Carl-Gustaf Rossby with the rotating tank used to study the motion of fluids in the atmosphere and ocean. Credit: NOAA/Department of Commerce.

west, in order suddenly to make a sharp turn towards the south, and attacks the warm air in the flank; it penetrates under it as a cold West wind."[2] These organized lines of clouds could be deduced from observations at regularly spaced intervals, then tracked as they moved across Britain, the Netherlands, and into the skies above Denmark, Sweden, and Norway.

Rossby was the man the U.S. Air Force trusted to bring the necessary meteorological know-how to the pilots in time for it to matter. To do that meant getting training programs (one would not be enough) up and running as soon as possible.[3] And to do that, as was so often the case during wartime, meant calling upon women to do jobs that they would not ordinarily be offered. So when Gerould went to see Rossby about the possibility of doing some meteorology courses alongside her astronomy degree, she ended up with an offer instead: to teach on the cadet training course that Rossby had established to quickly boost the military's forecasting capability. While she had not gone seeking such an opportunity, she was more than

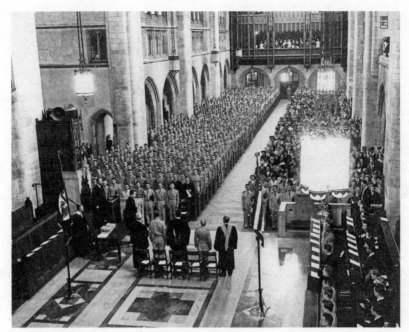

FIG. 5.2. Graduation and commissioning of U.S. Army Air Force meteorology cadets at the University of Chicago, September 6, 1943. Credit: University of Chicago Library, Special Collections Research Center.

ready for it when it arose. And while she had not yet, in fact, fallen in love with the study of the clouds, stepping into Rossby's office was a fateful step on her journey toward an intellectual passion that would last a lifetime.

Clouds, Gerould would later write, were more complicated than almost anything else. The only thing more complicated, she conceded, were human beings. "The mysteries of cloud formation, and the precipitation that can follow, have proven to be one of the most challenging aspects of the global climate system. Except for man himself, the weather is probably the most variable, unreliable, and fluctuatory phenomenon of which human intelligence has dared to attempt a science."[4] A cloud suffers the buffets of the atmosphere around it, retaining its shape for a while before becoming utterly changed. To become entrained, in meteorological terms, is to be taken up by a pre-existing air current or cloud. It is what happens to air in the environment that comes close to a cloud. "Within ten min-

utes, I was entrained in his orbit," was how Gerould described her
first meeting with Rossby.[5] It is no accident that Gerould used this
term, since she herself had used the concept it described (though
she did not invent it) to create a completely new way of thinking
about clouds and, as a consequence, a new way of thinking about
the circulation of the entire atmosphere.

Part of what Bjerknes and Rossby had begun, which Gerould
would continue, was the project of moving the study of clouds
beyond the scientific equivalent of stamp-collecting. By the 1930s,
when Rossby arrived in America, the study of meteorology had different ambitions. Those were twofold: first, and most urgently, to
provide operational support to the military in order to help pilots
make informed decisions about where and when to fly. Second,
Rossby and those who worked with him wanted to transform meteorology into a physical science. By this, they meant a science at the
heart of which lay physical equations that described the movements
of the atmosphere. There were obvious linkages between these two
desires, but they were also, perhaps to a surprising extent, distinct.
It was possible to make meteorological forecasts in the absence of
physical theory. And physical theories were not always that useful
when it came to making practical forecasts. It wasn't clear, then, in
which direction progress would first be achieved, and by whom.

Gerould taught for a year on the course, from the fall of 1943 to the
summer of 1944. That was enough. After that, she'd been entrained
by the study of meteorology. She then enrolled in a one-year master's program and continued to take classes after it ended. So it was
that she found herself still studying in 1947, listening to a series of
lectures on a topic that had been more or less ignored by the Scan-
dinavians who'd put together the modern science of frontal weather
systems: tropical meteorology. What Gerould heard electrified her
and caused her to abandon once and for all her ambivalence about
a field that she had good reason to doubt could ever provide her
with an all-important steady income and intellectually rewarding
work. The sense of an exciting new area of study opening up before
her eyes was almost palpable and too compelling to resist. "Almost

immediately," she later remembered, "a bolt of lightning struck me and I said to myself and my colleague, 'This is it—tropical cumuli are what I want to work on.'"[6]

The lecturer to whom Gerould owed this electric realization was Herbert Riehl, a man just eight years her senior. A Jew, he had been forced to flee Germany as a young boy, traveling first to England and then to America, where he himself became entrained by meteorology, almost by chance. He'd come to the United States with the idea of becoming a screenwriter and had pursued that passion for a few years. But success was not forthcoming and, seeking more practical employment, he applied to enter a U.S. Army Air Corps training program. The electrical engineering course he applied to was full, so he settled for meteorology. After completing the one-year program at NYU, he too went to see Rossby, and Rossby offered him the same opportunity he'd offered Gerould. Riehl accepted, and he taught on the training course at the University of Chicago the year before Joanne Gerould, from 1941 to 1942.

By 1942, the war in the Pacific had taken a dangerous turn. The Japanese had taken Burma, Malaysia, the Dutch East Indies, the Philippines, and Thailand. To combat the Japanese threat, military pilots desperately needed a much better understanding of the meteorology of the tropics. Thousands of military sorties over the tropical Pacific made it glaringly obvious that the weather worked very differently there than it did in Northern Europe. Sudden squalls arose in the absence of obvious fronts and demanded explanation. Rain fell from skies far too warm to sport ice crystals. The situation wasn't simply confusing; it was potentially dangerous. To fly safely, pilots needed better predictions of bad weather. The U.S. Army Air Corps agreed when Rossby proposed adding a special tropical institute to the nine-month cadet program. Arrangements were made as quickly as possible, and in the summer of 1943 the Institute for Tropical Meteorology in Puerto Rico was created in hopes that new observations and concerted effort might come up with something useable in time to help the war effort.

Riehl spent just two years in Puerto Rico, first as an instructor and then as director of the fledgling institute, before being sent

back to Chicago at the war's end. His time there was transforma-
tive. The meteorology that had been so proudly and confidently pioneered in Bergen was almost completely useless in the tropics. Tor
Bergeron's theory of rain formation, the pre-eminent such theory
of the Bergen school, required the presence of ice crystals. Without
ice, Bergeron had theorized, rain could not fall.[7] That may have been
true in Norway, but a single evening in Puerto Rico was enough
to demonstrate how patently false that theory was in the tropics.
Riehl vividly remembered such an evening, his first in Puerto Rico,
when "some of the staff walked along the beach, and admired the
beauty of the trade-cumuli in the moonlight. Well-schooled in the
ice-crystal theory of formation of rain, they had no suspicions about
clouds with tops near 8,000 feet where the temperature is higher
than +10 C. Suddenly, however, the landscape ahead of them began
to dim; then it disappeared; a roar approached as from rain hitting roof tops. When some minutes later they stood drenched on
a porch, drenched and shivering, they had realized that cloud tops
with temperatures below freezing were not needed for the production of heavy rain from trade-wind cumulus. There and then the
question arose: How is it with the other theories in so far as they
concern the tropics?"[8]

Back in Chicago, with his tropical experiences still fresh in his
mind, Riehl raised this question with the students sitting in front of
him. He described how at the end of the war, the navy had allowed
a small group of researchers based at Woods Hole Oceanographic
Institution (WHOI) to use some of their planes and ships to undertake some research on the trade winds of the North Atlantic. The
project was charmingly informal, an example of the kind of do-it-
yourself ethos that characterized WHOI at the time. Together, Jef-
fries Wyman, a physical chemist, and Al Woodcock, a self-taught
jack-of-all-trades, made some of the first measurements of temperature and velocities both inside and outside the so-called trade-wind
cumulus clouds (the clouds found in the region just north and south
of the equator where the winds blow consistently from east to west,
and toward the equator).[9] Their data put to rest forever the idea
that the tropical atmosphere was organized into fronts. Instead,

Woodcock and Wyman had demonstrated that the equatorial atmosphere displayed what one scientist later described as a "disconcerting sameness," with endless fields of trade-wind cumuli—isolated puffy clouds, like those in a children's storybook—stretching to the horizon.[10] This in itself was a strikingly different meteorological visage from that displayed by northern skies, where storms were common and clouds organized in long frontal systems. And there was more. Hidden within this seemingly calm atmosphere was the capacity for sudden, violent storms. Unlike in upper latitudes, when squalls arose in the tropics, they did so without any apparent provocation. Rarely, but unforgettably, they gave rise to a monster storm known in the Pacific as a typhoon and in the Atlantic as a hurricane. What caused these storms to arise when and where they did remained highly uncertain.

The data gathered by Woodcock and Wyman raised more questions than it provided answers. What caused the puffy trade-wind cumulus to form? Did the sea surface play a role in their formation? When and why did storms develop out of this seemingly uniform sea- and airscape? Like the atom, tropical clouds seemed to contain the hidden potential for dramatic transformation. The challenge was to explain what caused a seemingly harmless patch of tropical atmosphere to change into a violent squall and from there into an even more violent hurricane.

Listening to this news from a distant part of the planet, and the seemingly endless questions it raised about the basic workings of the atmosphere, Gerould had felt a rising sense of excitement, akin to an epiphany, that this was the work to which she wanted to devote herself. It remained far from clear whether that would be possible. In 1944, she had married a fellow University of Chicago student, Victor Starr, who had recently become the second person to be awarded a PhD in the University of Chicago meteorology program. Joanne Gerould had become Joanne Starr. She'd given birth to their son, David, in June, at the end of her master's degree. When she told Rossby of her plans to study tropical cumulus clouds, his response was cutting: "That's fine. An excellent problem for a little girl to work on because it is not very important and few people are interested in it, so you should be able to stand out if you work hard."[11] Undaunted, Starr immediately wrote to a friend of the family at Woods Hole, asking for a summer job. She got the job and spent the summer working on the Wyman and Woodcock cumulus data Riehl had lectured about.[12]

For Starr, it was an exceptionally busy time. Soon after David's birth, she started teaching physics to students at the Illinois Institute of Technology, spending summers at WHOI continuing her research on the cumulus cloud data. She convinced a somewhat reluctant Riehl, who claimed to know little more than she did about these clouds, to supervise her doctoral work. Given this level of activity, it is little wonder that something had to give. That something was her marriage. Joanne and Victor Starr were divorced in 1947, leaving Joanne with a young son to look after and the challenge of maintaining a research career that had barely begun. By now, she was deeply committed to a career in meteorology. Her chances of succeeding were not improved by her status as a divorced mother of a young child. In 1948, she married yet again within the circle of University of Chicago meteorology department, this time to Willem Malkus, a physicist studying for his PhD under Enrico Fermi. Now Joanne Starr Malkus, she received her PhD in 1949, and with it became the first woman in the nation to be awarded the advanced degree in meteorology. Another son, Steven, followed in 1950. All the while, she continued teaching at the Illinois Institute of Technology and traveling to Woods Hole in the summers to continue her research on cumulus clouds. Only in 1951 was she offered a paid research position—her first ever such job—at Woods Hole, which had already become a favorite location for both work and home life. At twenty-eight years old, she was returning to the skies she had left as a teenager just nine years earlier, with paid work as a research meteorologist.[13]

As a mother of two young children, Malkus might have chosen to continue doing the theoretical work on cloud models that she'd already begun. But she would never be satisfied simply analyzing other people's data, and in any case, there was too little of it to answer the questions she wanted to answer. Years later, she remem-

FIG. 5.3. Joanne Malkus analyzing data from the Pacific Cloud Hunt at Woods Hole Oceanographic Institution, with a long roll of cloud prints draped across the table. Credit: Schlesinger Library, Radcliffe Institute, Harvard University.

bered the moment that she realized she would need to do her own airborne studies, during a conversation with Henry Stommel, then a young oceanographer at Woods Hole:

> One day we were sitting there sort of talking at the blackboard and beating our heads around. You know we can't go any farther in this until we get some new observations. Why don't we see if the Navy still has any of those PBY aircraft and maybe we can not only put back the instruments we had in the Wyman expedition, but also make measurements of a few more things, particularly to get vertical velocities and liquid water . . . We sat around . . . saying "Do we really want to do this, are we willing to commit all the time

FIG. 5.4. Joanne Malkus in a DC-3 on a field trip to the Caribbean from Woods Hole Oceanographic Institution in 1956. Credit: Schlesinger Library, Radcliffe Institute, Harvard University.

to undertaking all the instrumentation of the aircraft, and installing the instruments and using screwdrivers, flight tests, calibration tests, and so on." We finally decided that we had to, there really wasn't any choice about it; that we were not going to get any farther understanding the physics of clouds with making models of clouds without making further observations and taking what we had learned from previous observations and models. . . . We went into it quite consciously, realizing it was going to eat up a big part of our lives. It was with a certain degree of ambivalence.[14]

She may have been ambivalent, but she did not remain still. She managed to gain access, as she had hoped, to an old navy airplane, and off she went, flying out from Woods Hole into the open skies and open waters south of Cape Cod. The closest tropical waters were near Bermuda, and that is where she headed. She was not alone. The airplane itself was kitted out with as many instruments as could be made to work on it and, in addition to the pilot, there was a photographer on board to help capture the clouds.

FIG. 5.5. Joanne Malkus loading instrumentation aboard the Woods Hole Instrumental DC-3 for cloud flights over and near Bermuda, c. 1955, with colleague Andrew Bunker. Credit: Schlesinger Library, Radcliffe Institute, Harvard University.

The ride was noisy and bumpy, but it was noisier than it was bumpy, and so Malkus and the photographer communicated by written note. She began, "The first run we got should be pretty valuable (fingers crossed) despite other subsequent difficulty. The nose camera contribution will be a vital part—because I do think we did get in to the most active part of the bubble and the film will show that—so things could be one whale of a lot worse!!" The response followed, written just below: "But how about the lens being 'less dry' (courtesy J. S. M.) that it sees nothing but droplets? Ah! The misery of this life (joking, we are in the beautiful tropical atmosphere)." And Joanne came back again: "Silly creature—it didn't get wet until we first went in the cloud did it???" And received the following response: "Yeah! Yeah, but the PBY didn't bounce either until we got inside, or at least not much until then."[15]

The notes are full of acronyms and banter. J. S. M. is of course Joanne Starr Malkus. The beautiful tropical atmosphere refers to the skies around Bermuda. The PBY is an amphibious plane developed by the navy for use during the war. Attached to it were a number of devices, including a nose camera for recording the size and location of the clouds, as well as a set of instruments for measuring temperature, humidity, and density of the clouds and the surrounding atmosphere as the aircraft stitched its way into and out of the cloud, observing it at a range of altitudes. The plan was to study these clouds in order to better understand how an apparently calm atmosphere could give way periodically to violent storms.[16] Flying in and out of the same cloud five or six times, Malkus and her crew performed the seemingly impossible task of fixing a cloud, rendering permanent the evanescent collection of water droplets. Without the airplane and, more specifically, the instrumented airplane, she would never have been able to achieve her objective. Key were the decisive movements of the airplane itself, which traveled not quickly, as might have been expected to capture the evanescent cloud forms, but slowly, to lessen the impact of aircraft speed on the measurements.

The notes describing the challenge of keeping the lens dry and the airplane stable survive because they recorded another similarly evanescent phenomenon—the burgeoning relationship between Malkus and the photographer, a man she referred to only as "C.," even fifty years later. Malkus kept these notes for the rest of her life because they captured a fleeting moment that mattered deeply to her, the burgeoning moments of a relationship that was to become one of the most important of her life.

FIG. 5.6. Joanne Malkus with the crew of her first research aircraft, which was on loan from the navy to Woods Hole Oceanographic Institution. Credit: Schlesinger Library, Radcliffe Institute, Harvard University.

The first time she'd seen C., she'd felt an instant attraction. It was "truly love at first sight in my case," she recalled in 1996. "This emotion is still strong 52 years later, 15 years after his death."[17] But this was not a conventional love story. In 1951, when these notes were made, Malkus was married to Willem Malkus. She had met C. when they found themselves working in the same institution, and soon, on the same project. Malkus had learned to see the atmosphere as a place whose tranquility belied a potential for rapid and dramatic change. And so it was with other people. With C., she learned how in an instant a relationship could shift from one of distance to breathtaking intimacy.

She probed her feelings in a diary she kept at the time with the same attention to detail and the same desire to follow an investigation to its logical conclusion that she demonstrated in her cloud studies. She wrote in pencil in a simple black-and-white ruled notebook and addressed her thoughts directly to C. "Why am I planning to write numerous letters to you, when it is highly unlikely that you will ever read them?"[18] Her answer to the question was that the diary entries could constitute half of an imaginary conversation with C. "By recording fragments of these," she writes, "I, at least, may learn something." In the same way, by observing a cloud from every angle, she hoped to learn "what makes the cumulus clouds grow, how they grow, what stops them from growing and the role they play in trapping moisture, heat and momentum."[19] For Malkus, learning about people and clouds was similar, requiring many observations taken at many angles. And just as clouds could only be understood in relation to their environments, people could really only be understood in relation to others.

One of the key scientific outcomes of the project was to prove that it was possible, using a slow-flying airplane, to gather useable data about the clouds. A more substantive conclusion, based on that data, was that it seemed to be the case that larger cumulus clouds were formed by the interaction and aggregation of smaller clouds.[20] It wasn't simply that small clouds grew into bigger clouds, in other words, but that big clouds were formed out of the groupings of smaller clouds. That meant that in order to understand clouds, it would be necessary to consider their interactions at multiple scales.

Malkus now began thinking about how and whether individual clouds and cloud behavior could be connected to larger-scale weather. What, she wanted to know, was the function played by small-scale convection—the movement of hot air—on larger-scale processes such as the movement of air from the tropics into higher latitudes?[21] In 1954, she used money from a grant she received to travel to the UK, where she presented her findings and sat in on lectures on cloud physics and precipitation at Imperial College with the aim of establishing "exchange of ideas and persons" in order to bring about "the vitally needed merging of the fields of cloud dynamics and cloud physics."

Malkus was not alone in wondering about the relationship between scales ranging from the molecular to the planetary or in finding inspiration in Woodcock and Wyman's data.[22] The first glimpses of the complexity of the tropical atmosphere had also fired the imagination of Henry Stommel, then twenty-seven years old and looking for good problems to work on. He wrote his first scientific paper on entrainment, presenting the then-controversial and counterintuitive idea that it was impossible to separate the study of clouds from the study of their surroundings.[23] In the mid-1950s, the entire field of meteorology was grappling with the question of scale, some of which had been raised by Stommel's paper on entrainment.[24] Much as oceanographers had once focused on the Gulf Stream as a phenomenon separate from the basin in which it occurred, meteorologists had long focused on clouds as discrete objects that could be analyzed separately from their surroundings. It was becoming apparent that it would never be possible to understand parts of the atmosphere in isolation. Only by looking at the overall circulation could individual parts be truly understood. Or, as Victor Starr had put it, "attempts to formulate *ad hoc* explanations for individual details of the general circulation without due cognizance of their role as functioning parts of a global scheme" would be doomed to failure. Something more was needed: an appreciation of the total meteorological picture. Meteorologists wanted to know how what happened within clouds affected what happened in the massive storms known as cyclones or anti-cyclones, and how such storms themselves related to the so-called general circulation of the atmosphere. What connections and feedbacks existed, and where did the discontinuities lie? This was a daunting proposition, but in 1951, Starr approvingly noted a new focus on the "essential oneness of the atmosphere which must be studied as an internally integrated and coordinated unit."[25]

The biggest reason for this change in the kinds of questions meteorologists were asking was the amount of new data becoming available. The airplane was essential, but another airborne device—the radiosonde—proved just as important. It consisted of a hanging basket of meteorological devices connected to a weather balloon that could transmit data on temperature, humidity, and pressure via radio to a receiver on the ground.[26] With radiosondes and airplanes, meteorologists could soar up to 30,000 feet into the atmosphere. It was now possible to imagine a global meteorology in which the

FIG. 5.7. U.S. Army Air Force meteorologists prepare to launch a hydrogen-filled balloon with a radio-sonde that measured temperature, humidity, and pressure. Credit: NOAA Photo Library.

motions of the entire atmosphere of the entire planet—along both vertical and horizontal dimensions—might be observed. No longer would meteorology be bound simply to a thin slice of atmosphere at ground level or to a specific region, as the Bergen school and the Institute of Tropical Meteorology had been. To transform global data into a global science, however, more than just observations were needed. Both novel theories and new ways of manipulating data were also needed to create what Rossby called, in the title of a landmark 1941 article, the "scientific basis of modern meteorology."[27]

In addition to the airplane and the radiosonde, there was one other great new meteorological instrument of the postwar era which would come to be essential for Malkus, as it would be for nearly every other working meteorologist. By 1946, its moment had arrived. In that year, the *New York Times* revealed plans for a "new electronic calculator, reported to have astounding potentialities."[28] The machine, measuring some eighteen by twenty feet long, would be capable of performing the "the most incredibly complicated and advanced equations in inconceivably minute fractions of a second." Though the super-calculator had been initially conceived as a tool for calculating the trajectories of ballistic missiles, almost immediately its meteorological potential came to the fore. John von Neumann, a professor at Princeton and the leading theorizer—and promoter—of electronic computing, argued that it could have "a revolutionary effect" on weather forecasting. These new machines were especially suited to repeating the same set of operations on an ever-changing set of data, precisely the kinds of calculations that were needed to solve the "nonlinear, interactive and difficult" problems that faced those trying to predict the weather.[29]

For those who had read Richardson's 1922 paper imagining the processing power of 64,000 human computers, it seemed as if the future had finally arrived. But while Richardson had dreamed only of forecasting the weather, the prospect of controlling weather and even climate was both an exciting and a potentially troubling new twist. The very first news report on the planned supercomputer noted that not only would it soon be possible to forecast the weather more accurately than ever before: It might even make it possible to "do something about the weather."[30] From the start, the purpose of the weather-calculating supercomputer would be to indicate not only likely future weather, "but also the points at which fairly small amounts of energy could be applied to control the weather."[31] The super-calculator, in other words, was always, at least theoretically, a weather-control machine.

Though von Neumann passionately believed in the redemptive possibilities of computing, he understood that fear was as important as hope in generating support for the project. Weather and climate control was a classic dual-use technology. In the right hands, it could lead to the alleviation of drought and famine, safer aviation, and even the improvement of climate for leisure and enjoyment. But it could also be used to wreak havoc on previously unimaginable scales. "Present awful possibilities of nuclear warfare may give way to others even more awful," he warned. "After global climate control becomes possible, perhaps all our present involvements will seem simple."[32] This moment of control, simultaneously feared and anticipated, seemed imminent. Not only was computer power sure to identify the necessary triggers, but the scale of the technological intervention that would be needed to affect climate at the global scale was no greater, von Neumann estimated, than that which had built the railway and other major industries.[33]

Just as a small nudge could send a boulder caroming down a mountain, relatively small inputs of energy could work on the atmosphere to produce massive effects. "The pull of a trigger is enough to release the energy in an enormous mass of air," explained a reporter in the *New York Times*. "Pull the trigger at the right place and we could ride the whirlwind and divert it to regions where it can do no harm."[34,35] A hurricane could potentially be diverted by igniting oil in key locations. Rain could be summoned by sprinkling coal dust on land to absorb heat. The details remained to be worked out, but already in 1947 it seemed clear that "the weather makers of the future are the inventors of calculating machines."[36]

For all the visceral horror and Promethean ambition such climate fantasies provoked, the computer was not only a tool for world-making or -unmaking. It was also a cerebral device that had the potential to extend the realms of thought—rather than action—in previously unimaginable directions. Once brute calculations could be organized along scientific principles, the computer became a tool for thinking about the atmosphere.[37] As such, it had the potential to transform meteorology into an experimental science. Not only could the computer enable the sorts of direct modification of weather or climate that could serve as experiments, but something more novel would become possible—a new kind of meteorological thought experiment, also known (with quotes in the original) as a "weather model." In this way, the computer enabled experiments to be done on a controlled atmosphere, safely removed from the realm of geopolitics where any atmospheric experiments raised special concern in the wake of Hiroshima and Nagasaki. "Nothing in plaster or wood," as an early commentator clarified, "but something that lies more in the mind and on the plotting board." This mental space enabled "an assumed earth" to be shaped according to the questions "we wish to ask of it, with an increasingly complex imaginary earth slowly built up of the constituent parts we add to it, a simple ocean, a series of rudimentary mountain ranges, a certain amount of water vapor." Thanks to the understanding such models facilitated, "we can begin to think of making weather to order on a regional scale."[38] If the model reproduced observed phenomena, it was a good indication the science was on the right track, "just as the birth of a child who resembles a paternal grandfather legitimizes both itself and its father."[39]

Such "imaginary" uses were implicit in the early plans for the application of electronic computing to numerical weather prediction. Weather forecasts—which is what the computers were initially conceived of being able to do—are, after all, imagined futures. The difference between numerical weather forecasts and so-called "weather models" is that models were intended to be used as tools for understanding the weather processes, while forecasts were usually addressing much more immediate and practical questions.

In Woods Hole, Malkus was applying these new ideas and new computing power to the tricky task of describing the growth of individual clouds. Using data she had gathered in the bumpy PBY flights, she created the first numerical cloud model that described, using a series of physical equations, how clouds grew and developed.[40] This was groundbreaking work—the first such "model" to attempt to reduce cloud growth to a series of equations. But it was only a start. Her studies of individual clouds only deepened her curiosity about how the process of convection acted on the larger scale. Both the temporal and spatial scale of regional or even global atmospheric motions required much more computing power than was available. And even if the computing power did become available, the problem remained much too complex to solve fully using the brute force of numerical computation. Better physical understanding was needed before more complex models could be contemplated.

Malkus took another tack. First, she convinced Herbert Riehl, her former supervisor, to join her as a partner on the project. Together, they started to look at data not just about tropical clouds. Instead, they looked at a vastly increased scale, the entire tropical zone, which stretched ten degrees on either side of the equator all the way around the planet. This was a scale of investigation that had previously been impossible. Now, thanks to the data streaming in from airplanes and radiosondes and plotted on global maps, Malkus and Riehl were able to form a clearer picture of how the atmosphere was moving around the entire planet—and to identify a gaping hole in theories of the general circulation. They identified this hole by tracking the movement of the sun's energy around the planet and in the process stumbling on an unexplained gap in that transfer— like a missing participant in a game of telephone. Somehow energy was moving around the planet, but the details remained fuzzy about where and how.

The sun is the source of all the energy on our planet. When its rays hit the earth, the angle and shape of the planet determine how much light is received by different parts of the planet. At latitudes higher than thirty-eight degrees in both hemispheres, the earth loses heat. Only between thirty-eight degrees and the equator—roughly the latitudes occupied by the African continent—is the radiation balance positive. But the planet generally maintains its average temperature at a fairly stable level. So the planet as a whole must act to transfer heat from the region around the equator to the poles; otherwise it would begin to cool off. To complicate matters, sea-level winds at the equator—the so-called trade winds upon which sailors had long depended—blow very consistently toward the equator. While it was generally accepted that the heat was carried aloft over the equatorial regions, and then transported poleward at higher altitudes, it was unclear what the precise mechanism for this transfer was. Somehow heat must be getting from the surface of the equatorial ocean— where the heat so efficiently absorbed by the water was in turn radiated upward—to the higher levels of the atmosphere known as the troposphere where the winds blew toward the poles. But measurements had shown that the middle layers of the atmosphere—

between the surface of the ocean and the troposphere—did not have nearly enough energy to transfer the heat upward. The middle layers were a kind of dead zone, energetically speaking. This left a mystery. How was the hot air getting from the sea surface up to the troposphere?

In addition to the "weather models" in which an increasingly complex earth was constructed out of equations representing physical phenomena, a new breed of studies had been developing since 1920 or so.[41] It was to these studies that Simpson and Riehl now turned. Called bookkeeping studies, they worked on the principle that in order to understand the planet, it was sometimes advisable to (momentarily) set physics aside. Just as an accountant processes transactions in order to balance a business's account books, so too was it possible to "process" the heat in the earth's climate in order to reach a certain balance, or equilibrium. In these studies, all that mattered was the increase or decrease of a chosen variable, be it heat, angular momentum, carbon dioxide, or any number of other quantities (such as, for example, ice, ozone, tritium, methane, and sulfur).

As suggestive as these papers were about the role played by small-scale phenomena such as eddies in larger-scale atmospheric features, no one had yet considered that cumulus clouds could play a role in large-scale circulations. This was the task Malkus and Riehl now set themselves. Their guess, based on their study of the radiosonde data and a rather bold intuitive leap made in the absence of other evidence, was that the heat was traveling upward from the ocean's surface in narrow regions of exceptionally buoyant air. These columns, or "hot towers," in which water vapor was condensed into droplets and released its heat, were the "overgrown brothers" of ordinary trade-wind cumulus clouds. They were giant—commonly reaching up to 35,000 feet high and sometimes as high as 50,000 feet—but relatively sparse. At any given moment, there might be only a few thousand active across the entire planet. These could serve as escalators for an enormous amount of heat, which was thereby able to bypass the lower layers of the atmosphere in which the winds were blowing back toward the equator. The lopsidedness

of the situation struck Malkus and Riehl forcefully. "The most striking conclusion from this work," they summarized, was the fact that
"only about 1500–5000 active giant clouds are needed to maintain
the heat budget of the equatorial trough and thus, implicitly, to provide for much of its poleward energy transport!"[42]

The hypothesis—and it remained only a hypothesis, with little
direct evidence to support it—solved the mystery of how heat from
the surface of the tropical oceans could be transported high enough
in the atmosphere to be carried on winds blowing away from the
equator. It also linked the energetics of the ocean and the atmosphere in a way that very few meteorologists (or oceanographers)
had yet done. Hot towers showed that the atmospheric circulation
could only be understood in relation to the ocean that supplied most
of its heat. Clouds could play an outsized role in the climate system, just as the impresarios of climate control had imagined. Even
without proof and a rather wide degree of uncertainty, the hypothesis was suggestive enough that Malkus and Riehl did not hesitate
to publish.[43] Cumulonimbus clouds of the necessary height—some
40,000 to 50,000 feet—had been observed. The question remained
whether there were enough of them transporting enough heat to
resolve the paradox. They ended their paper calling for more observations, during the upcoming International Geophysical Year, to
enable their theory to be refined and tested.

The challenge for Malkus and Riehl, and for others, was to generate not separate meteorologies—of fronts, of hot towers, of the
tropics, and of cyclones—but a single science that could, somehow,
describe the linkages between these scales. The theory of hot towers had seemed to solve the mystery of how energy was transferred
from lower to higher altitudes in the tropics, but it had created
another. How would it be possible to characterize (or understand) a
"system" in which large-scale regularities—the general circulation—
were in part determined by the most seemingly ephemeral and
fickle of phenomena?

"What we were doing that no one else had ever done before
was to put in the cloud systems as a key part of tropical energetics," explained Malkus, "and thus have energy move up scale in the

circulation size hierarchy, rather than down as in classical hydrody-namics."[44] She was keenly aware of how strange, and theoretically complex, such a system was. "It is no small wonder," wrote Malkus, "that the global circulation system operates in fits and starts, with its evanescent cylinders, of transient numbers, whose very exis-tence depends upon the vagaries of the flow itself!"[45] Malkus and Riehl had put their finger on a weather trigger, just the sort of thing the promoters of weather control were after. But what good was an evanescent, difficult-to-locate trigger that seemed, somewhat para-doxically, to depend for its existence on the large-scale phenomena that it might be said to affect? Here was a confoundingly circular and dynamic world of multiple scales that seemed to lack any reassuring hierarchy. The prospect of control seemed remote.

Malkus and Riehl's first impulse, after publishing this suggestive paper, was to do some more observing. Given the complex relation-ship between wildly disparate scales, they thought the only way to make further advances would be to "study these widely different scales of motion in their context to each other." They set out to make the "first attempt, largely descriptive, to relate synoptic and cloud scale phenomena." With this research program, Malkus was mov-ing a step toward her goal of linking small-scale phenomena, such as clouds, with larger-scale phenomena such as storms, hurricanes, and finally the general circulation of the entire atmosphere. It was an exciting time. The long era during which the cry had always been "we need more observations" seemed to be finally coming to a close, as airplanes and radiosondes began gathering more data in more places than ever before.[46] Drawing on the Wyman and Woodcock trade-wind expedition, Stommel's 1947 entrainment paper, and a series of other papers demonstrating how important the surround-ing atmosphere was in the formation of clouds, Malkus and Riehl summarized their findings in a book titled *Cloud Structure and Dis-tributions over the Tropical Pacific Ocean*. In it, they demonstrated why it was no longer be possible to look at the tropics and see a boring, steady-state atmosphere. Henceforth, the tropics would be seen as a tumultuous place, far more variable than it was stable.[47] Rain fell in the tropics far more erratically than anyone had imag-

FIG. 5.8. Joanne and Herbert Riehl puzzling over hurricanes, in a cartoon by Margaret LeMone. Note the "real hand-analyzed data" and the big question: "Why are there so few hurricanes?" Credit: Margaret LeMone.

ined. In regions where the majority of rain fell on just two or three days a month and even annual averages varied significantly, averages were not merely unhelpful but actively misleading.[48]

Underlying this optimism, for those who cared to notice, was a groundswell of doubt and uncertainty. It was one thing to have observations and the means to make calculations based upon them, but would "mere" observations ever really be enough to crack the atmospheric code? Physical insights, not just observations, were required to reduce an otherwise potential deluge of data into a usable current. "It is only through the leaven of some purely physical hypothesis," cautioned Victor Starr, "that we are guided to the appropriate mathematical use of these principles."[49] Where to find that leaven? The most useful tool in the scientific arsenal for reducing complexity was the experiment, a controlled intervention that enabled a researcher to isolate and test aspects of an otherwise overwhelmingly complex problem. Computers had raised the possibility of identifying likely points for experimental intervention. But the ability to perform controlled physical (rather than computer) experiments in which certain variables were held stable while others were manipulated had long eluded atmospheric scientists, partly for the reasons that Victor Starr had underlined. The atmosphere was so big, so unruly, and so "essentially one" that it was almost impossible to render it a pliable experimental subject.

Clouds could be—and had been—reproduced in the laboratory, including memorably by John Tyndall himself, but these miniature artificial clouds failed to capture all of the salient features of natural clouds. The motions of fluids more generally had been fruitfully investigated by Dave Fultz in a laboratory at the University of Chicago beginning in 1950. There he'd set up what were affectionately called rotating dishpan experiments. By heating a round tank of water, rotating it, and then dropping dyes into it, Fultz captured pictures of changes in the flow that reproduced some of the large-scale features of the general circulation of the atmosphere and ocean, such as the jet stream and other atmospheric waves. Using this apparatus, Fultz and others were able to reproduce some atmospheric phenomena artificially.[50]

The laboratory work by Fultz and others was useful but also frustrating, precisely because of how important scale was to matters both oceanographic and atmospheric. Much could be learned from reducing the ocean or atmosphere to a dishpan-sized model, but much was inevitably missed from such a set-up. The only way to truly understand the atmosphere, many felt, would be to experiment on it directly. The idea of an atmospheric experiment was almost unavoidable in these years, following a war that had been brought to a close by a grand and terrible atmospheric experiment that had produced, in the skies over Hiroshima and Nagasaki, an entirely new cloud.

As darkly potent as the radioactive clouds released by atomic weapons were, other, less obviously powerful technologies made surprising and important contributions to the growing sense that experimenting on the planet was not only inevitable but a necessary part of the progress of human knowledge. It was specifically the domestic freezer, a new appliance designed by General Electric to meet the growing demand of America's housewives for convenient and nutritious food with which to feed the postwar baby boom, which heralded a transformation in meteorological practice.

In 1946, in the laboratories of GE, a young engineer named Vincent Schaefer had been playing around with creating supercooled clouds inside one of these consumer freezers. After generating clouds made up of the supercooled water vapor expelled from his lungs, he experimented with dropping bits of dry ice into them. Immediately, and dramatically, the clouds precipitated into snow. His colleague Irving Langmuir predicted that atmospheric clouds found outside GE freezers would respond in the same way. Bernard Vonnegut (brother of author Kurt) then demonstrated that silver iodide could be a very effective cloud seeder (more effective, per gram, than dry ice). In 1946, Schaefer succeeded for the first time in seeding a cloud *in situ* with dry ice. It was the beginning of a bonanza of cloud seeding, in which states across America (mainly in the dry Western states) sought to solve their agricultural worries with the expeditious application of a few kilograms of silver iodide.

In 1947, under the auspices of Project Cirrus, Langmuir seeded the first hurricane using this technique. The effects were disastrous. The storm, which had been headed northeast over the Atlantic off the coasts of Florida and Georgia, abruptly reversed track and headed west, making landfall in Georgia and South Carolina. Though the observers on-board the aircraft which had seeded the storm did not measure any changes in the structure or intensity of the storm (which might have indicated that the seeding had been the cause of the change in its direction), Langmuir nevertheless could not resist claiming "success" in this instance, even though the landfall had resulted in damage.[51] The local towns sued, and cloud seeding was flagged not as a source of knowledge but one of potentially limitless liability.

Such episodes demonstrated how strong was the desire to exploit what remained a little-understood aspect of cloud physics—the role played by seeds, or nucleators, in prompting precipitation. Bernard Vonnegut's brother, Kurt, was inspired by these events to write *Cat's Cradle*. Ice-9, Kurt Vonnegut's imaginary corollary to silver iodide, turned everything it touched not to water, but to ice. The consequences were terrible, and the message of the tale was as clear as the destructive ice: Interfere with the dynamics of nature at your peril.

The distance between visionary dreams and inadvertent consequences was shorter than most imagined. In 1957, Roger Revelle and Hans Suess published an article in which they described the widespread emission of carbon dioxide via the burning of fossil fuels as a "large scale geophysical experiment."[52] This now-famous sentence is often presented as a prescient call to arms, one of the first to alert humanity to the risks of an uncontrolled intervention into the planet's climate system. Revelle and Suess did emphasize the novelty of the situation, noting that this experiment "could not have happened in the past nor be reproduced in the future." But rather than warning of the danger of unchecked emissions, Revelle and Suess were urging their fellow scientists to take advantage of an unprecedented opportunity to study the ocean, much as Rossby had mused on the possibility of covering the polar caps with coal. They used the term *experiment* in the classical sense of a scientific test designed to eliminate as much uncertainty as possible. "This experiment, if adequately documented, may yield a far-reaching insight into the processes determining weather and climate." Careful measurement and observation, in other words, could transform a merely unwitting (and uncontrolled) intervention into a proper scientific experiment. Revelle and Suess accordingly urged, as Malkus and Riehl had, that data be collected during the International Geophysical Year which could be used to track the path taken by this excess carbon dioxide as it traveled through the "atmosphere, the oceans, the biosphere and the lithosphere."[53]

Ever on the lookout for opportunities to do more observations, Malkus quickly realized that hurricane studies could be a continuation of her cloud studies by other means. A new opportunity presented itself in the wake of a series of natural disasters. In 1954 and 1955, a series of harsh hurricanes pummeled the East Coast of the United States. In quick succession, hurricanes Carol, Edna, Hazel, Connie, and Ione battered the coast, destroying more than six billion dollars' worth of property (in 1983 dollars) and killing nearly 400 people. In response, Congress appropriated funds for a National Hurricane Research Project (NHRP), to be headed by Robert Simpson, a meteorologist who had been a forecaster during the war and had helped set up a wartime weather school in Panama. Intervention was written into the plans for this government laboratory just as it had been for the first supercomputers.[54] The mission was explicitly tasked with studying how to modify hurricanes artificially, along with more basic research into the formation, the structure and dynamics, and the means of improvement of forecasts of hurricanes. The new funds meant airplanes and airplanes meant government scientists could now, for the first time, do *in situ* cloud studies on tropical clouds that stretched from the surface of the ocean all the way up to the troposphere.

Malkus saw that the NHRP could be a platform for a more genuinely experimental research program into the link between cloud and atmospheric dynamics she had been hoping to pursue. In 1956, she flew to Miami, where she met Bob Simpson for the first time. Here, finally, was a chance to help turn meteorology into an unambiguously experimental field science, with the rigor and attention to documentation and control that had been missing from most previous cloud-seeding projects.

While the NHRP was established in an attempt to distinguish

hurricane research from the seat-of-the-pants, under-theorized, and overhyped work done by Schaefer, it was impossible to start with a clean slate. The memory of the hurricane that had slammed into Georgia possibly as a result of intervention was fresh, and when it came time to draw the boundaries within the Atlantic where hurricanes would be fair game for intervention, an excess of caution was applied. The result was that only one or two hurricanes a season passed through the area in which seeding was allowed.

Still, the chance to use the new instrumented aircraft funded by the NHRP was too good for Malkus to pass up, and while she had focused on clouds up until that point, she didn't see the sense in making distinctions between what were obviously related phenomena. "So I thought, well gee, I'd better get into this too. Hurricanes are, after all, systems of tropical clouds. Systems of tropical clouds that somehow get together and run wild. Why did they happen in that way?"[55] She started reading up on hurricanes, and soon she had come up with an idea that linked the hot-tower hypothesis she and Riehl had developed with the formation of hurricanes.

She was fascinated in particular by the "calm central eye, surrounded by furious winds." What, she wondered, explained this phenomenon? Relatively little was known about hurricanes, because radiosonde and aircraft sounding were scant. She pored over the data there was, including a film made by MIT which pioneered the use of weather radar to probe the eyewall of the 1954 hurricane Edna. Looking carefully at the film, she realized that most of the air in the hurricane eye came from the cloudy eyewall.[56] With Riehl, she developed a model of how a hurricane develops which emphasized the importance of the ocean as an "extra" heat source.[57]

At the same time that Malkus was using hot towers to think about hurricane formation, Robert Simpson had begun developing his own theory about how to modify hurricanes. He thought that if you could seed certain key clouds (equivalent to hot towers) in the eyewall, then you could force it to re-form farther out in the storm, thereby reducing the intensity of the wind and weakening the storm.

On September 16, 1961, Robert Simpson was able to test his theory when a naval aircraft dropped eight canisters of silver iodide into the eyewall of Hurricane Esther. Instead of continuing to grow as it had been, the storm maintained a constant intensity. Thanks to the coordinated observations of crew aboard six airplanes monitoring the storm, the response of the storm to the seeding was recorded in great detail. These synchronous radar observations showed that kinetic energy had been somewhat reduced in the eyewall. The next day, another load of canisters was dropped, but they missed the eyewall. Subsequent observations indicated that the storm had retained the same intensity it had after seeding the day before. From the difference in the storm's evolution in response to seeding and no seeding on successive days, they deduced that the seeding had been successful. In an article for *Scientific American*, the two researchers wrote with no small measure of pride that instead of "merely observing" the formation of a hurricane, they had attempted to "interfere in a critical area with the delicately balanced forces that sustain a mature hurricane." They took some pains to point out the novelty of the work, noting that their experiments were among the few "ever performed on an atmospheric phenomenon larger than a single cumulus cloud." Despite the potential risks, there were good reasons to experiment with hurricanes, not all of which involved modification. Better forecasting seemed an almost guaranteed outcome once hurricane research had graduated from an "observational discipline to an experimental one."

But forecasting was only the start. Hurricanes were the perfect place to test assumptions about weather and climate triggers. Precisely because they are so massive, any attempt to modify one will fail unless it is precisely targeted. Failure to modify, then, would prove a theory's limitations, while successful modification would mean the theory was likely correct. For this reason, attempts at hurricane modification seemed like ideal tests of hurricane theory, with the bonus that if the theory proved good enough and a hurricane could be precisely targeted, truly staggering amounts of energy would be within human control. In practice, it was extremely difficult to know if in fact an intervention had been successful. If you don't know what it would have done otherwise, how can you know if you have changed a hurricane's behavior?

There was a paradox here. Successful intervention required precisely the advanced understanding of hurricanes that such an intervention was designed to help generate. Despite this limitation, those in charge of government funding deemed the seeding of Hurricane Esther a success. Soon after, a new project was established which was explicitly, and solely, aimed at modifying hurricanes: Project Stormfury, founded in 1962 as a joint undertaking of the U.S. Navy and Department of Commerce. Malkus's numerical cloud models were critical to the justification of the project, providing a tool for testing assumptions and generating predictions against which modification attempts could be checked.

Malkus herself had mixed feelings about weather modification. Though she was attracted by the research possibilities and the more distant potential for humanitarian applications, she was also wary of the corners that were often cut when it came to cloud seeding. Asked in 1961 to comment on the potential for hurricanes to be diverted, she said, "I wouldn't say we're on the threshold, but weather control is not a totally ludicrous idea."[58] The problem was that interventions were often staged "with too many claims, with an underestimation of the enormous natural variability of the system, and with impatience on the part of the management to get a positive result in a short period of time."[59]

Despite her doubts, two things convinced her to come on board as an advisor. The project was relatively inexpensive and had potentially huge benefits to humanity. And, just as importantly, Stormfury offered a way to improve her models and learn more about hurricanes. "I believed the Stormfury Project would be the only way I could do the experiments on cumulus clouds which I had been thinking about for some time."[60] Rather than trying to seed hurricanes in order to modify their courses, Simpson saw seeding as a tool for doing experiments in the atmosphere. "People should be placing their emphasis on weather modification as atmospheric experiments and I've said so all along." While it was feasible to change the development of individual clouds with seeding, the modification of hurricanes with the intention of benefiting humankind was, she thought, always a "very long shot."

With this in mind, Malkus signed on to Project Stormfury. The plan was to bend the practical goals of project to her own scientific aims—to make of modification both a scientific tool and a practical intervention. She would never otherwise be able to muster the number of aircraft necessary to monitor an experiment well enough to determine if it had been successful.[61] In 1963, she got exactly what she wanted, when she carried out a seeding experiment that "changed my life and that of many others."[62] Stationed in Puerto Rico in mid-August of that year, Malkus and the rest of the Stormfury team were waiting for Hurricane Beulah to develop an eye well-formed enough to be modifiable. In the lull before the storm, Malkus saw a chance to test her ideas about cloud growth.[63]

During the experiment, a total of six aircraft and several dozen technicians enabled Malkus to make successful measurements on eleven non-hurricane clouds, of which six were seeded and five were controls. "When that first cloud exploded," she remembered, "I was never more excited in my life."[64] The scientists and crew of the several aircraft also broke into wild celebrations on seeing the growth. All save one of the seeded clouds grew explosively, while the control clouds did not. The results were just as Malkus's model had predicted. She had managed to do what she had long hoped for—to use seeding as a tool for atmospheric experimentation, and to get the full force of the navy's aircraft backing her as she did so.

Malkus and Simpson published the results of their cloud-seeding efforts in *Science*, and the magazine put a dramatic series of pictures of the exploding clouds on its cover in the summer of 1964. The response from the public was instant and intense. It was, in Malkus's words, "an immense storm" for which neither of them was prepared. The "intensely interesting effects" that had been produced in the seeded clouds stoked hopes and fears that the time for weather control had finally arrived. Some greeted the arrival of a hoped-for utopia of weather control, while others saw a repeat of the hubristic meddling with nature that had led to the bomb.

As exciting as it had been to watch the seeded clouds surge

FIG. 5.9. Cover of *Science* for August 7, 1964, illustrating the results of Joanne Malkus and Robert Simpson's cloud-seeding experiment.

upward, Malkus and Simpson were careful to make it clear that the most significant outcome of the experiment was not the explosive growth but the demonstration that the experiment itself was possible. They wrote an article for *Scientific American* explaining the nature of the cloud-seeding experiments and trying to pin down the meaning of control. On the one hand, the seeding had shown that "now a real atmospheric phenomenon is at last subject to a relatively controlled and theoretically modelled experiment." It was true, they thought, that clouds could finally be turned into experimental subjects. But the kind of control needed for scientific experiment was preliminary to—and less complete than—that needed to be able to manipulate hurricanes to human ends. That kind of control—what Malkus and Simpson called "real control"—would be longer in coming. Rather than a giant leap forward, they cautioned that "here meteorology is taking the first small steps toward becoming an experimental science, which it must become if man is ever to exert real control on this atmosphere."[65]

The navy and the Department of Commerce were not interested in theoretical models but in modifying real hurricanes. The same weather system that had enabled Malkus and Simpson to test their model using cloud seeding also proved amenable to more practically oriented interventions. Just a few days after the successful cloud seeding, Hurricane Beulah had obligingly developed a more mature eyewall. The entire hurricane—not just a nearby cloud—was now ready for seeding. Using many more aircraft and significantly more silver iodide to massively seed the eyewall, the navy pulled out the stops to see if modification was possible. On the first day that seeding was attempted, the special silver iodide bombs missed the eyewall and no effects were seen. The next day conditions for seeding had improved, and this time the bombs hit their target. Measurements of the core of the storm showed that the pressure dropped precipitously following the second seeding and the cloud pattern of the storm changed dramatically, with the eyewall dissipating and forming ten miles farther away from the center of the storm, much as Malkus and Simpson had predicted.

Despite the seeming success, it was impossible to say on the basis of one modification attempt whether the seeding had definitely caused the changes to the hurricane. The natural fluctuations of hurricanes were so big, and the nature of cloud patterns so little understood, that much remained unknown. Repeating the experiment was one way to test the hypothesis, but given how much these storms varied naturally, it could take centuries to "separate statistically the man-made changes from the large natural fluctuations."[66]

In 1964, the National Academy of Science (NAS) convened a panel on weather modification to provide advice on how best to

proceed in an area that was both scientifically and ethically challenging. Malkus was a member of the panel, along with Jule Charney, Ed Teller, Ed Lorenz, Joe Smagorinsky, and others. The panel
cautioned against haste and noted that evidence to support the
efficacy of seeding remained thin. There was as yet no data to suggest, for example, that so-called winter orographic storms, such as
those in Colorado which were the subject of great interest on the
part of farmers and ranchers in the state, could be made to produce
significantly more rain, nor that hurricanes could be steered, nor
that black dust or other surface coverings could produce rain. The
evidence did not exist to support a leap into weather modification
on an operational basis. Most present efforts were characterized by
a "seed first, analyze later" approach from which very little reliable
information could be gleaned. Patience, counseled the panel, was
needed. It could take decades, not years, before the physics was
well-enough understood to support widespread weather control.
A split developed between research scientists and state legislators
of arid states, some of whom accused the scientists of being more
interested in producing papers than water.[67] Much remained unresolved, even as experts like Malkus gave good reasons to proceed
cautiously. In the same year that the NAS panel advised restraint,
Congress passed a special resolution appropriating $1 million for
operational weather modification programs.

The first small steps toward "real control" of the weather which
Malkus and Simpson described in 1964 are today reminiscent of
those taken most famously by Neil Armstrong five years later as he
stepped onto the moon. But it was to another momentous speech
that the scientists may have been referring. That speech, given by
President John F. Kennedy on July 26, 1963, occurred just weeks
before the cloud-seeding experiments were carried out. In the tele-
vised address, Kennedy, speaking in a measured voice in a tone of
somber hope, announced the partial ban on nuclear tests in the
atmosphere, ocean, and outer space that he and Soviet statesman
Nikita Khrushchev had been able to engineer after years of difficult negotiations. Kennedy called the agreement a "shaft of light"
in a time otherwise characterized by suspicion and tension, and "an

important first step—a step toward peace—a step toward reason—a step away from war." He reiterated the metaphor in the closing lines of his speech, ending with what felt like an audacious hope: "and if that journey is a thousand miles, or even more, let history record that we, in this land, at this time, took the first step."[68]

For Malkus, life changed dramatically in 1965. She had left Woods Hole in 1961 to take a position at the University of California, Los Angeles. That same year, she gave birth to Karen, her daughter with Willem. In the meantime, her relationship with Bob Simpson, which had developed over the course of the shared work at the NHRP and on Project Stormfury, turned into something deeper. The cloud-seeding experiments and the modification of Beulah gave rise to what Malkus called the "Malkus/Simpson collaboration and increasing close friendship."[69] In 1964, Malkus divorced Willem Malkus and left her tenured position at UCLA for a research position at the U.S. Weather Bureau. That career move, an unlikely one at first glance, was necessary because nepotism laws prevented a husband and wife from working at the same institution. On January 6, 1965, Joanne and Bob Simpson were married. With the wedding, Joanne took the name of Joanne Simpson and the directorship of Project Stormfury, at the Weather Bureau. Thus began what she called her second great love, and a partnership of mind and spirit that would last until her death.

If Joanne Simpson had finally found contentment in her personal life, the controversy over weather modification raged on. In 1963, she had celebrated in the skies above Puerto Rico as she watched the seeded clouds explode upward, and shared in the sense that hurricane modification was possible when Beulah seemed to respond to seeding. The Stormfury hypothesis that seeding the supercooled water around the eyewall of a hurricane could cause it to release latent heat and migrate outward, weakening the storm, seemed to be correct. These early days of optimism and excitement turned out to be misplaced. Circumstances would conspire to make it impossible to carry out the research necessary to determine whether the hypothesis was, indeed, correct. As difficult as it was to coordinate six or even ten aircraft flying through a hurricane, it turned out to be much more difficult to create a statistically powerful enough program of experiments to tame the natural variability of these great storms. Hurricanes are extremely variable objects. In order to understand their motions—both natural and modified—it is necessary to study a lot of them. This is always an expensive undertaking, and sometimes an impossible one. From 1963 until 1968, no eligible storms passed through the experimental area. Meanwhile, the controversy over weather modification continued. By 1967, Joanne Simpson, no longer willing to put up with the tension surrounding the program, resigned. Stormfury continued, with uneven but ultimately unconvincing results. When, in 1969, Hurricane Debbie finally obliged the researchers and five seeding runs were made, the results were deemed consistent with a revised Stormfury hypothesis (which required less instability in the eyewall and deployed massive, repeated seeding just outside the eyewall instead). But eligible hurricanes that passed through the safe zone remained frustratingly rare, making it impossible to further verify the eyewall hypothesis. Over the course of the 1970s, research into hurricane modification tailed off, and when eventually the project was canceled in 1983, it was deemed a failure.

There were to be no easy answers to the question of whether modification could be done, much less whether it *should* be done. Nevertheless, weather modification was, already, a reality. Both intentional and inadvertent modification of the atmosphere was already happening, in places both near and far. In order to have any hope of distinguishing the artificial state of the atmosphere from its natural state, basic atmospheric processes would need to be understood. Simpson and her fellow panel members noted (in 1964) that the major barrier to figuring out which clouds could be seeded was the "great natural variability" of clouds themselves, which included differences in drop sizes, water content, ice content, temperature structure, internal circulation, and electrification.[70] This natural variability made careful statistical evaluation both necessary and "very difficult."

What was very difficult on the small scale was both vexing and potentially catastrophic on the larger scale. Though it was not yet possible to "induce perturbations to trigger massive atmospheric reactions," such a day might be foreseeable. What was not yet within reach was the capacity to predict the effects of such a major modification with "continent scale or larger" extent. Until that was possible, the committee concluded that "*to embark on any vast experiment in the atmosphere would amount to gross irresponsibility*" (italics in original).[71]

In order to understand the effects of artificial modifications to weather and climate, the committee suggested that a theory of natural climatic change was needed. Instead of performing experiments in the atmosphere, where their effects would be hard to interpret and which could have unintended consequences, the committee suggested the safe space of the computer model within which "the consequences of artificial perturbations could be assessed."[72] The boundary between the earth and the air, and the sea and the air, was a crucial and understudied aspect of what they still called the large-scale atmospheric circulation (rather than the coupled ocean atmosphere). Field experiments here seemed feasible and much needed, along with numerical studies.

The panel included a section on inadvertent atmospheric modification, a problem with no sign of abating. "We are just now beginning to realize that the atmosphere is not a dump of unlimited capacity, but we do not yet know what the atmosphere's capacity is or how it might be measured."[73] The committee also noted that the pollution caused by cities was capable of affecting the local climate. As Rossby and Revelle had, they remarked on the scientific potential of such a "continuing experiment in climate modification."

When Simpson left the Weather Bureau, she returned to academia, taking a job as professor of atmospheric science at the University of Miami and taking on the directorship of an experimental meteorology laboratory at Coral Gables. There she continued the work she had started on Stormfury, convinced that if she applied her dynamic seeding technique to small-scale cloud structures, she would be able to generate a testable hypothesis about the potential for generating rainfall from seeded clouds. But this project, too, was to be stymied. If individual clouds could be seeded with her technique, she wondered if seeding could cause more so-called cloud mergers, groups of clouds that were naturally productive of the most rainfall in Florida. Simpson calculated that she would need to do several hundred seeding experiments to detect a fifteen percent increase in rain in the target area. Her boss was unwilling to fund even a hundred cases. Thus the pattern set in Stormfury of research projects handicapped by a lack of data continued. Simpson felt that one of the fundamental assumptions of Stormfury—that supercooled water was present in abundance in hurricanes—had been somewhat slanderously called into question. According to her, data gathered by the NHRP that demonstrated the presence of such water was later thrown away, and new results suggesting no such water was present were used to justify the cancellation of the project.[74]

Simpson's personal involvement in weather modification—both at Stormfury and on the Florida experiments—was a source of regret for her. Weather modification for her had always been a means to an end. That end was not the transformation of clouds or storms, but the gathering of data. She deeply regretted the cancellation of the weather modification programs even as she had distanced herself from them. In a speech she gave as American Meteorological Society President on October 4, 1989, she remarked on the bitter irony that many meteorologists had celebrated the demise of weather modification programs—which they viewed as unscientific—since it was the cloud physics community which suffered from the demise of weather modification research the most, as "badly needed new observational data on clouds is much slower and harder to come by than in the heyday of weather modification experiments."[75]

Simpson's own thirst for new data sent her toward the last great project of her long career. She moved to a new Laboratory for Atmospheres at NASA's Goddard Space Flight Center, and in 1986 became the lead for the science team in charge of a Tropical Rainfall Measuring Mission. This satellite was the first of its kind, carrying a space-based rain radar that could peer deep into the heart of the clouds through which Joanne had so painstakingly flown. She worked on this project for eleven years before it finally launched. The satellite exceeded the goals the NASA scientists had set for it five years after its launch. In 2002, it measured the profile of latent heat released by tropical systems, providing a space-based confirmation of the work she and Riehl had done some fifty years before.

Joanne Simpson participated actively in the preparation of her archive for deposit at the Schlesinger Library at Harvard University. She annotated hundreds of photographs and wrote numerous short essays to accompany documents from different parts of her life. Much of the archive relates to her long and active career. She also decided to deposit some very personal documents, including the notes and journal she kept during her relationship with C. She explained her decision to share this intimate material in a letter to the archivist. "My work life is well known but I have deliberately kept my personal life as private as possible and hence if I should die before I finish, the material I am starting to send you now on my personal life would be lost, as little of it is known by anyone else."[76] She decided to forfeit the privacy she had guarded for a lifetime because she believed it was critical to portray the full complexity of her life in science.

As a woman in an otherwise almost entirely male profession, her personal choices had always been subject to public scrutiny—at least those personal choices that she could not hide from outside view. News stories breathlessly reported on her ability to cook and keep a house as well as maintain her professional career.[77] Her bold habit of flying through the clouds—and even hurricanes—was all the more surprising since Simpson was, by one writer's estimation, "a rather wispy and timid-looking blonde," who, in addition to being one of the top five meteorologists in the world, "runs a big home in Woods Hole, Mass, does all the cooking for her husband and two sons."[78] Despite the sexism, the journalists who wrote these pieces were correct when they noted how entangled Simpson's home and work lives were. Each of her three marriages was to a man who had emerged from the same small sphere of research meteorology, and her relationship with C. was likewise rooted in their shared experience of scientific work. To deny the centrality of these relationships to her life would be to miss something important. Simpson's decision to make this intensely personal material available to scholars was taken deliberately. She wanted to make it possible for the full story of her life, in all its complexity, to be told. And she hoped that a time would come when the so-called work/life balance was no longer seen to be a problem only for women, but for all working people.[79]

For this to happen, the archives will need to reflect the reality of life as it is lived. For now, the evidence of the multiple roles played by male scientists as fathers, husbands, and lovers remains frustratingly hard to come by. In Simpson's archive, a fuller portrait emerges, of a woman who lived intensely, and indeed passionately, throughout her long and productive life. If she chose to conceal much of the storminess of her private life during her life, in the afterlife she envisioned for herself the true complexity of her life was finally given space to breathe. Like the great clouds she studied, Simpson had allowed herself to grow expansively, sometimes with dramatic speed, into areas that had been presumed to be off limits. Her passions and her science were inseparable. "I think I am generally perceived as a pretty cool character," wrote Simpson. "Nothing could be farther from the truth. To understand how a woman, or a man, for that matter, creates original work in any field, it is necessary to penetrate the emotional masks, and my masks have intentionally been hard to penetrate."[80]

6

FAST WATER

At the age of twenty-seven, Henry Stommel was adrift. As a young scientist, he had a strong, intuitive sense of the importance of choosing good problems, but he didn't yet know what the right problem—scientifically speaking—might be. On the advice of a colleague at Woods Hole Oceanographic Institution, where he worked, he read a paper on hydrodynamics, the study of how water moves. This was a welcome piece of pure science, absent the military objectives that had dominated wartime work and which had disturbed him. It was a promising start. Soon after, at a dance hall in New York, he was introduced to Carl-Gustaf Rossby, who had established the department of meteorology at the University of Chicago (and met a young Joanne Gerould) during the war. This meeting was more decisive, a nudge in a direction he might otherwise not have traveled.

The world was small enough that the dance hall meeting resulted in an invitation to spend a semester in Chicago at Rossby's lab. He listened to Rossby lecture in a style that appealed to him. Though Stommel was uncomfortable with simple certainties, he was not afraid of simplicity itself. Rossby made bold physical simplifications in his drive to understand atmospheric motions, and this boldness appealed to Stommel. It may have had something to do with a chance event in his life. Stommel himself thought so. As a teenager, due to a typographical error, he received a prescription for eyeglasses that were much too strong for him. He could not read or

FIG. 6.1. Henry Stommel thinking with pen and paper, c. 1950. Photo by Jan Hahn. © Woods Hole Oceanographic Institution.

see the blackboard easily and learned to compensate by seeking out problems with relatively few components, problems that he could easily ponder with what he called his mind's eye.

The problem Stommel now decided to think about was why the major currents in the world's oceans are asymmetrical. Simple as it sounds, no one had yet thought to ask this question, though the phenomenon had long been observed. In every major ocean basin of the globe, the currents are stronger on the western than the eastern side.[1] This fact holds true in the Atlantic, the Pacific, and the Indian Ocean even though those basins have strikingly different coastlines and ocean floor landscapes. Topography could not account for it, so what, wondered Stommel, could? Remembering Rossby's boldness, Stommel imagined an ocean that was rectangular, straighter and simpler even than a bathtub. He then perturbed it only with a few

FIG. 6.2. Henry Stommel was a maker, here hammering something c. 1950. Photo by Jan Hahn. © Woods Hole Oceanographic Institution.

variables—wind stress at the top and friction at the bottom—and added in the effect of the rotation of the earth on the waters within. He painstakingly hand-calculated the effects of these simple variables on his even simpler ocean, completing his calculations with a slide rule. He discovered, to his surprise, that this simple model of the ocean reproduced the crowding of streamlines on the west. He wrote this up in a paper that was five pages long. It was called "The Westward Intensification of Wind-Driven Ocean Currents."[2]

He was not yet twenty-eight years old, and he had just created a new science, dynamical oceanography. It was concerned with understanding how the waters of the oceans move. What Stommel had shown is that it is possible to describe the large-scale movements of water in the ocean using physics and mathematics. He had done so without a PhD, a fact about which he was for some time self-conscious, despite the advice he received from elder oceanographic statesman Columbus Iselin, director of WHOI. He'd written to Iselin asking for advice about whether he should pursue an advanced degree. Iselin replied that "If you are set in making a professional career in the geophysical sciences, I doubt that a PhD is worth a cent to you. In so far as it would take you time to earn, it would, in fact, cost you good money."[3] The movements of the ocean could be, at least in certain respects, deduced from very simple physical laws. This is how Stommel described it, many years later: "There is a great hydrodynamical machinery of the ocean" that "governs how the flow of the water responds to the winds that drive it at the surface and to the differences of density that are sustained by climate at different latitudes."[4] In other words, a good mechanic can hope to understand what it is that makes the oceans move.

Stommel believed in the existence of this machinery. He believed it was possible to describe the ocean according to the laws of fluids in motion. But—and this is an important if rather a subtle point—he didn't believe that insights into the oceans could be deduced from such laws. The laws were too general, and the ocean too complex. He believed that the only way to achieve understanding was by a leap of insight followed by a process of exhaustive iteration. He called the leap a "seed image," something that he had to invent from somewhere that lay beyond analysis, beyond even language. The iteration is the relationship between this image and something Stommel called reality, for which he took observations of the ocean to be an acceptable proxy. He saw this as a process of crystallization, or, more accurately, of trying (and mostly failing) to get crystals to form. Once he accumulated enough ideas about a particular problem, but had achieved no great insight, he went into an almost trancelike state. He described this state in his memoir. In order to achieve insight, he wrote, "I defocus my mind, to deliberately lose it all, to melt the fragments of ideas into something akin to a hallucinatory vision. In effect, I try to raise the conceptual temperature to some equilibrium value where structure disappears for a few days, and then try lowering it to see what crystallises out." Rarely was one round enough. Once an image was achieved, he then tested it against observations of the actual ocean. What he wanted to know was whether the image from his mind helped generate the right kind of ocean, the ocean that was actually observable and whose motions it was possible to measure.[5]

Henry Stommel had grown up near the sea, in Brooklyn and on Long Island. He had studied mathematics and astronomy as an undergraduate at Yale, and when he graduated in 1942, he had registered for the draft as a conscientious objector. Despite his objections to war, he was assigned to teach the mathematics needed for navigation to other young men being prepared by the navy's accelerated V-12 training program to serve as officers, a task he ended up enjoying despite its military setting. When the war ended, he enrolled in the divinity school at Yale, but he soon realized that he was no more comfortable with the certainties offered by religion than he was with those offered by war.[6]

Then, in 1944 he took a job as a researcher at WHOI. His undergraduate education and his experience as a teacher on the navy's training course had equipped him enough to get him a job at what was then a small but fast-growing institution. He did varied bits of research, but nothing really captured his attention. He was adrift. And there he might have remained, but he was lucky (he was fond of saying so himself). He had come of age at a time and a place when the public was interested in sponsoring scientific research. It was also, not incidentally, a time when the military was pouring vast amounts of money and resources into the sciences that could tell them how to safely fly and land airplanes and how to hide (and detect) submarines at sea. He was lucky, most immediately, to be surrounded by others who took a benevolent interest in him.

Appropriately enough, his first address in Woods Hole was the old rectory house for a local church, a generous building with plenty of room for a rotating assortment of live-in bachelor oceanographers. On the walk from home to work, Stommel could see the harbor, and he checked each morning to see whether the boats were still safe at their moorings. Life was both small and large, occupied by domestic amusements (of a bachelor flavor) and outsized adventures on the waters of the Atlantic that teased the shores of Cape Cod. The house in which he lived was itself a kind of ship. In it, the lines between work and life were blurred. A prankish atmosphere, full of punning and physical humor, reigned. This was a new kind of freedom: the conjunction of like-minded but differently skilled souls, each a specialist in his own area, sharing a passion for the ocean.

He also spent time on real ships, on the seas around Woods Hole and beyond. By his own admission he was an unadventurous sailor and a rather inept seaboard technician, trying to take temperature readings aboard a small ship in the icy winter seas off the Gulf of Maine, mostly in vain. He was not sure what the measurements were for, and he was not confident that they were accurate. The bathy-thermograph, the instrument then used for taking the water temperature, was itself intemperate, full of mechanical faults. Stommel tried mostly to stay out of the way of the men and the heavy instruments they deployed over the side of the ship. Despite his inelegance, he loved both the idea of being there and, for as long as he could stomach it, the being there as well. He developed a belief, which stayed with him throughout his life, that to know the ocean required spending time on it. He believed it was possible to develop a physical intuition for the water, a gut sense of how it moves, even on scales incommensurate with human experience. Seagoing was also a social activity; he got practice in working with many different kinds of people, and getting along with most of them. "Work at sea rubs off the sharp edges, and makes us better people," he later wrote.[7]

Woods Hole in those years was a little utopia of scientific research. Money from the Office for Naval Research (ONR) poured in, the legacy of oceanography's great wartime contribution and testament to the need to maintain expertise in the ensuing Cold War. During short trips out to sea, Stommel studied the upper layers of the ocean, trying to work out the patterns of hot and cold that enable submarines to hide in the acoustical shadows made when water bends sound. The pacifist in Stommel flinched at the thought of contributing directly to violence, but he saw the practical need for knowledge that could help in defense of his country, and the funds offered by the ONR in aid of a basic physical understanding of the ocean came with few strings attached. To refuse the opportunity to learn more about the ocean would have been perverse.

With the results of his five-page paper on wind-driven currents in mind, Stommel traveled to Britain in 1948, mimeographed copy of the paper in hand, to see what he could learn by observing rather than theorizing the ocean.[8] While his mind's eye had led him to imagine the effect of the rotation of the earth on the movements of a planet's worth of water, he was also driven to understand the disorder in the ocean. For this, he needed to start on a considerably smaller scale. He set out to knock on the door (this is almost, but not literally, true) of a man who had studied the behavior of fluids when they break down into turbulence at a range of scales, from the smoke that rose from chimney pots to seeds set loose on the wind and balloons that rose above excited crowds in Hyde Park and Brighton, fluttering tags which members of the public were encouraged to send in once the balloons eventually fell to earth. From these sorts of observations, this man, Lewis Fry Richardson, had come up with a deceptively simple equation that described the speed at which objects in a turbulent fluid separate.

Richardson's interest in turbulence arose in tandem with his dream—dating to the early 1920s—of achieving what he called "weather prediction by numerical process"; in other words, of predicting the future by crunching numbers. To do so meant coming up with mathematical equations that described the motion of the atmosphere. This included accounting somehow for the ways in which the movement of air was obstructed, and thereby disordered, through contact with vegetation and mountains at the surface, and through the collision of masses of different kinds of air—hot, cold, wet, and dry. Richardson was aware of the importance of atmospheric turbulence even as he recognized that to make any kind of weather forecast, he would have to dramatically simplify the weather of the planet, to make it fit into imaginary squares which he used to order the globe. Within these squares, 200 kilometers to a side, all messy turbulent phenomena were reduced to a single number.[9]

Some thirty years after Richardson had first envisioned numerical weather prediction as a far-off dream, Stommel traveled to visit him. Richardson had devoted the latter part of his life to understanding not natural systems, but human ones. A pacifist like Stommel, he had tried to use mathematics to explain why arms races happen, and why they unfold the way they do. This he considered more important than understanding how the turbulent fluids of the planet behave. But he had consented to join Stommel for a late return back into the nature of physical, rather than human, systems. Previously, Richardson's observations had been limited to the motion of the atmosphere. The question Stommel and Richardson both wanted to answer now was: How similar is the ocean to the atmosphere?

They did a watery version of the experiments on diffusion in the air that Richardson had performed two decades earlier using balloons and other objects. Stommel would remember for years to come the damp, heavy soil in the garden from which, on the suggestion of the older man, he dug up the parsnips—parsnips!—and the cold room in which he and Richardson worked together, slicing and weighing the vegetables, before they cycled down to the loch together. There, they walked to the end of the pier and dropped them into the water and tracked the speed at which they separated, using an ad hoc bit of equipment Richardson had worked up out of bits of wood and string to provide some means of accurately gauging the distance traveled by the vegetables.

Their choice of parsnip as experimental apparatus is telling. If you drop a parsnip into the ocean, its buoyancy is such that it will float low in the water, with just a sliver of itself protruding above the waterline, unaffected by the wind. Floating freely and easy to see, humble parsnips make good oceanographic measuring devices. When given the choice of a simple, robust approach versus a complex, relatively fragile one, Richardson—and Stommel—both chose the former. Later in life, Stommel frequently commented on his own perceived lack of mathematical skill and the effect it had on his working practices. "When I contemplate the superb skills of some of my colleagues as mathematicians, as instrument designers, as masters at squeezing information out of masses of data, as scholars

FIG. 6.3. Henry Stommel, "oracle" mathematician. Though he was self-conscious about what he felt were the limits of his mathematical skills, he credited them with forcing him to simplify problems. Photo by Jan Hahn. © Woods Hole Oceanographic Institution.

of encyclopedic knowledge, as scientific administrators with considerable power of decision, I realize how limited and amateurish my own ideas are," he explained. "Therefore when I get an idea, I simply have to pass it on to someone else who has the skills to develop it. That's not really generosity—it is just being practical."[10] Though he could be defensive about his limited mathematics, he also welcomed the way it forced him to simplify the questions he asked—as well as to seek collaboration with others.

Richardson was less conflicted. He spent his life seeing beyond the limitations that hemmed others in. The power of his imagination outstripped, by far, the power of any contemporary computational abilities when he imagined a great human machine for numerical weather prediction. He estimated that some 64,000 computers—then the only such kind were human beings—would be necessary to make it work, but such detail did not daunt him. He imagined something extraordinary which would eventually come to pass: a means of predicting the future state of the atmosphere based on a knowledge of its current state and a finite set of equations that would describe the motions of its particles.

He recognized that some features of the weather would need further study. It was not yet possible, he knew, to reduce all the features of the atmosphere to simple equations. He therefore envisioned experiments that would be ongoing—in the basement of the great meteorological theater—to study the motions of eddies, large spirals of whirling water that spun out from and eventually pinched out of major ocean currents, such as the Gulf Stream. Turbulence was too important, and too fascinating, to ignore. At the same time, it was possible—even necessary—to begin the task of crunching the numbers before waiting for turbulence to be fully understood. Richardson set about understanding turbulence by observing it, first with balloons and trails of smoke, then with thought experiments, and finally with Stommel at the end of the pier.

Together, the two men dropped forty-five pairs of parsnips off the end of the pier, and watched which way they went, hoping to understand what happened as they traveled farther away from each other. Based on what they observed of the bobbing parsnips in the Scottish loch, they concluded that energy diffused in the loch water according to the same principle as it did in the atmosphere. The results they achieved recall a paper Richardson had published nearly thirty years earlier, in 1920, describing the counterintuitive possibility that eddies acted like "thermodynamic engines in a gravitating atmosphere," which added, rather than dissipated, energy from the system.[11] The paper they published together is today remembered as much for the peculiar audacity of its first line—"We have observed the relative motion of two floating pieces of parsnip"—as for its conclusion that the atmosphere and the ocean exhibit similar forms of turbulent diffusion.[12] It is important to note that scale mattered to the final result. What happens in a bathtub's worth of water differs significantly from what happens in the loch, and what happens in the ocean again would require an even larger leap.

It was Stommel's Gulf Stream paper, rather than the parsnip paper, that would galvanize the field. It stimulated new work on that most familiar of ocean currents, work that would lead in turn to further understanding of the circulation of all the water in the oceans. The questions raised by Stommel's parsnip collaboration with Richardson about the role of turbulence in ocean circulation would have to wait considerably longer until they could be addressed. They lay in wait, aspects of the movements of water that could neither be solved nor ignored, like a shadowy creature whose exact dimensions are not understood, and which is only glimpsed in fragments. In the meantime, Stommel's mind's eye ranged freely across scales, considering the basin-wide gyre of water of which the Gulf Stream is a mere component and the relatively tiny motions that force those parsnips now this way, now that. It would take time, years and even decades, but eventually those twinned images of the ocean, at the very large and the very small scale, would be brought together once again, not only in Stommel's mind's eye, but in the minds of his fellow oceanographers. Then the seemingly quixotic attempt to map the motion of parsnips on a Scottish loch would be revealed as a step in the global understanding of oceans. For now, this all lay in the future.

Pressures in the ocean are almost unimaginably great. They are the reason that the depths of the ocean remain nearly as hostile and unfamiliar a place as the surface of the moon. Just ten meters of water provides the equivalent of an entire atmosphere's worth of pressure. The pressure at two kilometers below sea level is two hundred times greater than that of the atmosphere at sea level. This pressure is also the reason that it took so long for the discipline of oceanography to catch up with what any individual sailor knows in the gut: that water moves quickly and in ways that are somehow both ordered and chaotic. Experienced sailors know which currents run where, and what kinds of winds are to be expected in a given region, and they also know that the sea is a surprising and fast-changing place. What they know relates to but cannot directly access what lies below. For that, both instruments and ideas are needed which can connect a ship at the roof of the ocean to the mysteries that lie below.

For a long time, an ocean congruent with the felt experiences of sailors—an ocean that moved in the sometimes chaotic ways they reported—did not emerge from the descriptions of those who tried to study the oceans on the large scale. For most of human history, the ocean was accessed by the ship's sail and by a sounding line dropped over its side and a variety of instruments deployed at the end of it. Navigation on the surface of the ocean and penetration to its depths were possible but limited by the buffetings of wind and current. Bottles strewn on the surface gave a very rough idea of the speed of the very top layer of the ocean, but no instrument could go deep and stay there. Saltwater, pressure, strong currents, and wildlife conspired to render most instruments useless. Added to these problems was the basic difficulty of tracking and recovering an instrument that might be sent to flow with the currents. Under these circumstances, measuring the flow of deep water directly was impossible. Without the right tools, almost nothing was known about the depths of the ocean for most of human history. The result was that those who studied the oceans assumed that nothing of much import was happening deep below.

Despite these limitations, there were many aspects of the water that could be sampled. A key episode in the history of the study of ocean currents was the moment in 1751 when Henry Ellis, captain of an English slave ship, noticed that if he sent a bucket down deep enough when the ship was in warm waters near the equator, it was full of cold water when brought back to the surface. The only explanation for how such cold water could be found in such permanently warm places was that it had somehow traveled there from colder places—from the far north or the far south. In 1798, Benjamin Thompson (also known as Count Rumford) published an essay titled "Of the Propagation of Heat by Fluids," in which he noted that quite unlike freshwater, which begins expanding once it reaches four degrees Centigrade and continues to do so until it freezes, seawater contracts as it cools, right up to the point at which it freezes. Cold saltwater, Rumford saw, would be very dense, dense enough to sink to the depths of the ocean. The idea of a closed circulation—one that returned on itself—seemed to follow directly from the physics of freshwater. Rumford argued that the sinking of cold water in the ocean implied a circulation, or a current, consisting of the equator-ward flow of cold water and a corresponding and opposite flow at the surface.[13] The surface winds, which had so long been seen as the main force in moving the oceans, seemed to pale in significance to the great masses of water moving according to their density.

Much later, in the 1860s, William Carpenter, a physiologist search-ing for new species of crinoids (feathery echinoderms that were found to live at great depths in the ocean) in the North Atlantic, noted an area between the Shetland and Faroe Islands where warm and cold deep waters were found in close proximity. He developed a theory of what he called "general oceanic circulation," where the emphasis was squarely on the first word (to distinguish it from local circulation). His "magnificent generalization" was that the waters of the globe moved around the entire planet. Carpenter's idea was that cold water that sank at the poles was continually replacing the warm water that was transported north by currents like the Gulf Stream. Similar movements could be inferred in the Southern Hemisphere.

Not everyone agreed. James Croll, the Scottish autodidact who had come up with a grand theory to account for the ice ages, had strong opinions about the relative importance of wind or density to account for the movement of the oceans. His theory of the ice ages depended on the imbalances in the earth's climate brought about indirectly by very long-term changes in the eccentricity (or shape) of its orbit. For his theory to work, he needed wind to be a signifi-cant driver of ocean circulation. He argued that as ice built up at the poles as a result of feedback effects, the trade winds would also strengthen and therefore push the Gulf Stream farther to the north, adding even further to the cooling effect that had been triggered by changes in the earth's orbit. Due to insufficient evidence, Croll and Carpenter's disagreement about whether it was winds at the sur-face or the deep motion of dense water that held the key to ocean circulation reached a temporary stalemate.[14]

By the 1870s, mechanical instruments that could withstand the

monumental pressures at depth and the corrosive effects of seawater made it possible to accurately measure how warm, how salty,
and how deep any particular spot in the ocean was. In a laborious
series of surveys, tens of thousands of observations were made,
starting with the British expedition aboard the *Challenger* ship of
1872–1876. At the time the most expensive and comprehensive voyage ever undertaken for the purpose of studying the ocean alone,
the *Challenger* and her crew spent four years traveling 130,000 kilometers (70,000 nautical miles) around the globe. Fifty years later,
the German *Meteor* expedition made an even more systematic survey of a smaller portion of the world's oceans. Zigzagging between
South America and Africa fourteen times, they covered a similar
distance to the *Challenger*.[15] Soon, even the treacherous waters of
the southern ocean began to be measured, thanks to the efforts of
the British *Discovery* expeditions.

   Challenger, Meteor, and *Discovery*: the names say it all. These
were single-ship operations. No matter how long and ambitious
the journeys they made, the outcome of their expeditions was constrained by the simple fact of having been made by a sole ship. Measurements were by necessity taken one after another—serially—
and only after the expeditions finished were they were brought
together. By plotting the records of temperature and salinity on a
map, lines could be drawn between the dispersed data points. These
lines traced what were assumed to be the contours of actual bodies
of water in the ocean, masses of water defined as having the same
properties. But these maps performed a sleight of hand. By combining in a single image observations that had been made years and
sometimes decades apart, they created pictures of the ocean that
gave the appearance of being snapshots of the ocean at a particular moment. In fact, they were idealized—and, in a very real sense,
therefore imaginary—average bodies of water, based on measurements widely separated in both time and space.

   These atlases suggested the ocean was a very orderly place.
Great tongues of water were revealed to be spreading across the
ocean. By studying these slablike masses while keeping in mind
the basic properties of water—that cold water sinks and spreads

and that salty water is heavier than fresher water—it was possible to guess how the water might be flowing. These guesses were the hard-won product of thousands of hours of surveying. They gave specific contours to the theoretical movements of deep water that men such as Rumford and Carpenter had described in the nineteenth century. On the scales of time and space that these expeditions were able to measure, the ocean was a stable place in which certain large-scale features, such as the Gulf Stream and the world's other western boundary currents, stood out. Water in this averaged ocean acted more like cooling lava or even solid rock than a fast-moving liquid. There was no drama in these deep waters: nothing like the hurricanes and squalls whipped up on the surface, nothing equivalent to the thunderstorms, fronts, and cyclones of the atmosphere. Instead, there was the slow ooze of cold water as it moved across the ocean floor at a distinctly funereal pace—decades, hundreds, and even thousands of years were needed before much would change. Time passed slowly. Any features of the ocean smaller than several hundred kilometers across and shorter in duration than several hundred days could be (and were) easily missed from one data point to the next. The absence of these phenomena from the data led many to assume they must be absent from the oceans themselves.

This image of a slow and steady deep ocean was a product of inference. Taking the information that was available to them about the ocean—primarily the readings taken by the single-ship expeditions that had been compiled into atlases—oceanographers had used the basic rules by which physicists described the motion of fluids to infer the machinery of the ocean. It was impossible to do as Stommel did and iterate their inferences with reality, that is, with observations, simply because very few such observations existed. As late as 1954, a one-page table could easily list every such series of measurements ever made, the longest of which showed only the periodic changes associated with the tide and was therefore useless for understanding currents. The *Meteor* expedition had made its current measurements by dropping a meter overboard and attempting to keep the ship as still as possible above it—a tricky maneuver

that produced unreliable results. The data they collected had indicated some deep waters that moved more quickly than would be
expected based on the so-called dynamical method. But these measurements could still be explained according to the old paradigm
of a deep ocean that was, on average, sluggish. ("Even when deter-
minations of deep currents, based on several days' observations,
are not in agreement with the deductions by indirect methods, it
does not necessarily follow that either is wrong," wrote one analyst
of such data.)[16] Measurements at the surface sometimes revealed
small-scale features—swirls and eddies—which suggested movements that stood out boldly from the average currents expected in
the area, but these minor anomalies did not seem significant enough
to put the slow-motion entire ocean machinery into question.

Sometimes evidence emerged that seemed more forcefully to
contradict, or at least to complicate, this image of the deep ocean
as a lifeless and nearly motionless place. Sailors occasionally pulled
up from the depths strange and wonderful creatures that previous
theories of an abyssal oceanic dead zone had declared impossible,
feathered crinoids waving like *Alice in Wonderland* phantasma-
goria, barnacle-encrusted samples of the first telegraphic cables
to stretch across the Atlantic, witnesses to a strange underworld
which still had mysteries to surrender. Jarring bits of data were
also pulled up from the deep, when thermometers brought to the
surface showed warm water where cold was expected, salty when
fresh was in order, and vice versa. If plotted, these bits of data inter-
rupted the smooth contours of the water slabs, grains of grit in a
vast oceanic oyster shell. These anomalies were not entirely ignored
by oceanographers. Small-scale, random-seeming currents were
clearly represented on a map of Norwegian sea surface currents
that was created in 1909 by eminent Norwegian oceanographers
Bjorn Helland-Hansen and Fridtjof Nansen.[17] In hindsight these
messy squiggles seem important. At the time, they generally did
not. Often they were dismissed as mere noise in the data, a result
of kinks in the instruments, mistakes made by those who tended
to them. Sometimes they were viewed as accurate but simply irrelevant measurements. Such small-scale and supposedly short-lived

phenomena were assumed to have little effect on the larger circulation. The alternative—that small-scale phenomena might influence
large-scale circulation—was more or less unthinkable to oceanographers who observed the ocean at the time.

Theoreticians also had their reasons for believing that turbulence
on the small scale was unlikely to be a driver of ocean circulation.
Since Osborne Reynolds had first identified the moment at which a
flow of water transitioned from smooth to disordered motion, turbulence had been understood as a phenomenon by which energy
dissipated *out* of a system. In this view, turbulence was important
because it acted as a brake on the system, allowing energy to dissipate "down gradient," into ever smaller scales until the energy was
evenly dispersed throughout the system. It is this idea that Lewis
Fry Richardson captured in his memorable poem, included as a
frontispiece in his 1922 book setting out the possibility of numerical weather prediction. "Big whorls have little whorls which feed
on their velocity," summarized Richardson, "And little whorls have
lesser whorls and so on to viscosity."[18] The idea that there might
be significant interaction across disparate scales—that turbulence
might "skip" from big whorls to lesser whorls without stopping at
the middle whorls, or, more provocatively, that energy might travel
"upwards" from relatively small-scale motions in the ocean back
into the largest scales—was pretty much as anathema to theoreticians of fluid dynamics as it was, for different reasons, to seagoing
oceanographers.

Stommel knew very well the limitations of the data. For the moment,
there was little he could do directly about it. What he could do was
keep thinking about the ideas in his westward intensification paper.
Missing from his first stab at the problem was any acknowledgment
of how differences in the density of the deepest waters of the ocean
contribute to its circulation. Like others before him, he had concen-
trated only on the wind. Thinking about it more deeply, he realized
that the same physical logic that explains the crowding of streamlines at the surface of the western side of the ocean basins, stream-

lines caused by the action of wind, would also produce currents beneath them running in the opposite direction. He made the rarest of things: an oceanographic prediction. A southward current, never before detected, should be found beneath the northward-flowing Gulf Stream. His prediction relied on the forcing action of wind on the surface as well as the movements of deep water according to its density. For the first time, the forces that moved the water at its surface and in its deepest abysses—forces about which Carpenter and Croll reached an unbreakable stalemate in the 1860s—were joined in a single theory. Stommel united wind at the surface and differences in temperature and salinity in the deep to create a machine of ocean.

In doing so, he changed the way oceanographers think.[19] Instead of considering the Gulf Stream an independent and isolated oceanic phenomenon, similar to the flow of water gushing from a garden hose (which is how even Rossby thought of it), he imagined it as one part of a rotating gyre of water that spans the entire Atlantic basin. In a sense, what Stommel did is to show that the Gulf Stream doesn't exist separately from a basin-wide system, that it could only be fully understood as an aspect of a larger system. The payoff for this large-scale thinking was that the Gulf Stream—and the whole basin-wide system—became mathematically and physically explicable. But the cost of this insight was that henceforth the ocean would need to be considered as a whole. In revealing the hydro-machinery of the ocean, Stommel had also made a case for the interconnectedness of its parts.

His papers were both a challenge and a gift to those who cared to read them and recognize what they meant. Not everyone did, for they came in a deceptively modest package and spoke in a language that was, oceanographically speaking, new. For those who were listening, he had opened things up. Two things followed from these papers and the ideas behind them—one, a blossoming of theoretical interest in models of the ocean, and two, a series of expeditions designed to test these theories. These were related but separate developments. There was between them something of the iteration that Stommel had described as central to his own creative thinking.

The dance between ideas about the ocean and observations of it had sped up.

New pictures would be drawn, some like Stommel's—big and bold as a Rothko painting. Unlike the old atlases, the images that the theoreticians who approach the problem now came up with portrayed a much more active ocean. Water was moving not simply according to its density, but thanks largely to the effects of the wind. But for all their innovations, for the way forces were now directly enrolled in the quest to describe the oceans, these models were still constrained by the imaginations of those who created them. Those imaginations, fed on the data collected by mechanical devices (based in some cases on technology more than one hundred years old), still observed a sluggish ocean. So the theoreticians were bound to provide what was, in essence, a new explanation for an old description of what the ocean looks like. It was an ocean that was still sluggish, sticky and, to use the technical term, laminar: layered with neat slabs of water.

Thanks to Stommel's 1948 paper and the subsequent theoretical work that it inspired, the idea that the ocean was a moving fluid whose motions could, in theory, be explained by the physics of fluid dynamics had been established. But—and here was the rub—physics remained inadequate to deduce the motions of the ocean. The ocean was too big and too complex for the equations—and the computational capacity—of the time. (It remains, in many respects, too big and too complex even today.) Stommel's successful prediction of a deep boundary current is still the exception rather than the rule. New mental images, new oceanographic seeds, are usually generated not by the ferment of physical theories but by new observations. And those observations only become possible when clever people come up with new ways to measure the ocean.

The pictures of the ocean made by the observers in the years following Stommel's 1948 paper looked like pointillist portraits from which most of the dots had been erased. They were both (relatively) precise and fragmentary. They were made not out of the imagination but out of hours spent on ships and in workshops—making devices, tinkering with bits of metal and wire to see if something could be made robust enough to withstand the terrible pressure of the deep ocean and the pernicious effects of salt and water and yet remain sensitive enough to make useful measurements. These images, which seemed so limited to begin with, would eventually be the ones from which a whole new understanding of the ocean would arise. In 1950, just two years after the parsnip visit and the publication of Stommel's westward intensification paper, the first expedition designed to scrutinize the Gulf Stream using multiple ships at the same time got underway. This expedition marked the start of twenty years of exploration that would fundamentally transform our image of the ocean from a slowly moving slab of treacle to a turbulent fluid.

Like the Wyman/Woodcock expedition, the background of the two men responsible for the expedition (dubbed Operation Cabot) says a lot about the values it embodied. Fritz Fuglister was trained as a painter and was working as a muralist on WPA-sponsored projects on Cape Cod when he joined WHOI as a research assistant to work on drafting charts. He had even fewer formal oceanographic qualifications than Stommel did—precisely none—but that had not stopped him from bringing his intelligence to bear on the question of how to make the ship and its allied equipment a more powerful oceanographic instrument. Val Worthington was another so-called technician lacking advanced academic degrees.[20] In 1961, they would appoint themselves as members of SOSO, the Society of Subprofessional Oceanographers, in acknowledgment of their collective lack of professional degrees. (Stommel was the third and only additional member of this society.)

Using six ships and a set of recording devices that could measure pressure and temperature at greater depth than ever before, Fuglister and Worthington mapped the Gulf Stream over the course of ten days. They took temperature measurements simultaneously aboard the different ships, a coordinated effort that paid off when they happened upon a pronounced meander of the stream south of Halifax. They spent the final ten days of the expedition tracking this meander. As they did, they watched it extend south until it eventually pinched off from the Stream, forming a ring of fast-moving cold water—an eddy. It was the first time anyone had ever observed such a phenomenon unfolding in real time.[21]

Looking back, this moment appears to be a clear milestone—the first definitive sighting of an ocean eddy and therefore the "discovery" of weather in the deep ocean. The data Fuglister and Worthington gathered would eventually force oceanographers to consign their image of a sluggish ocean to the dustbin. At the time, the data was simply ambiguous. What Fuglister and Worthington had managed to do was record a single sighting of an elusive and still very mysterious phenomenon. Question marks festooned the results. Of these, the most pressing was to determine how representative a single eddy associated with the Gulf Stream was. It was possible such eddies were to be found only near powerful, fast-moving western boundary currents like the Gulf Stream. It was also possible, however, that the entire ocean could be laced with them. No one knew which was the case.

More data were needed to feed the iterative process of theorizing and observing. To collect that data, more and better instruments were required. So too was a big enough frame of reference. Fuglister and Worthington had relied on luck and a device called the smoked-glass-slide bathythermograph, a tool for recording temperature at depth, as well as the new system of long-range navigation based on radio waves, called loran. As lucky as they had gotten in finding and tracking their eddy, it still seemed almost impossible to measure currents at the depths and scale necessary to build up a good enough understanding to begin to relate site-specific measurements to a general theory of ocean circulation. Getting enough observations at deep enough locations spread out across the relevant area remained a major technological hurdle for which existing instruments were inadequate.

Stommel unleashed his imagination on the problem—envisioning a system of underwater devices that would operate like meteorological radiosondes, floating with the underwater currents the way such balloons float on currents of air.[22] Tracking the floats would depend on hearing them, and for this he imagined a timed explosion, a sort of underwater bomb that would alert listening devices to the location of the floats. This was an inventive but ungainly idea and, luckily, before he needed even to attempt to convince others of its practicability, Stommel discovered that someone else had come up with a simpler, more elegant solution to the problem. His name was John Swallow.[23] Using scavenged scaffolding tubes and a vat of caustic chemicals to thin them to precise thicknesses, Swallow was able to devise a tool sturdy enough to withstand the ocean depths but delicate enough to be adjusted (using ballast) so that it could achieve neutral buoyancy. Like an air balloon that could be made to hover at any altitude with the right combination of sandbags and hot air, Swallow imagined long floats that could be balanced to float in the ocean at predetermined depths. Instead of bombs he used a simple electronic circuit to create a ten-Hertz signal—a noise at a precise frequency—that could then be tracked by a pair of hydrophones on a nearby ship.

Swallow initially set up big tubs of water in the stairwell of the building where he worked, so he could carefully weigh and adjust his floats as necessary. By 1957, he was taking the floats out to sea. With Val Worthington, he went hunting for the deep countercurrent that Stommel had predicted would be found underneath the Gulf Stream.[24] The data Swallow managed to collect were not as clean as they needed to be to settle the matter. Even if the measurements indicated that there was a countercurrent below the Gulf Stream, it was impossible to know what this meant for global currents. The big question which Stommel and many others now wanted to answer was whether such motion could be observed in the depths of the mid-ocean, the places that were supposed to be the sleepiest, and most stable, of all.

Soon, more and better results were acquired. In the summer of 1958, Swallow went hunting again, this time for deep currents in the eastern North Atlantic off the coast of Portugal. He thought he could detect currents as slow as a millimeter a second, roughly the speed of the sluggish deep waters of the oceans.[25] Strange results began to appear as soon as he started to measure. Floats were found to move ten times faster than expected and to change directions abruptly. Two floats 2.5 kilometers deep and just 25 kilometers apart

FIG. 6.4. John Swallow preparing one of his neutrally buoyant floats for deployment, as the ship's
cat observes. Credit: Archives, National Oceanographic Library, National Oceanography Centre,
Southampton, UK.

moved at dramatically different speeds, one ten times faster than
the other.[26]

So strong was the prevailing belief that deep currents are weak
that even in the face of these measurements, a second expedition
was designed on the assumption that only slow-moving currents
will be found at depth. This is important because the ships that were
needed to track the floats could not refuel fast enough (they had to
return to harbor to do so) to find them if they went much faster than
one centimeter per second. In late 1959, using a ninety-three-foot
ship called the *Aries*, Swallow and others set out to study the mid-
ocean, in the Sargasso Sea west of Bermuda. The first readings that
they managed to take indicated something extremely surprising.
Expecting to find evidence once again for the deep northward flow
that Stommel had predicted, they found something else. It looked
like there were unexpected numbers of fast-moving eddies beneath
the surface, swirling vortices of water some 100 kilometers wide
with velocities some hundred times faster than expected. (Luckily,

the crew was able to change its approach to gathering data so that the ship could keep up with the floats.) Not only were the currents faster than expected, but they also seemed to increase in speed the farther down they were measured. Nothing in Stommel's theories had predicted, or could easily explain, such a result.[27] Not only had eddies been found in the north Atlantic, they were more powerful than had been imagined.

The tail was starting to wag the dog. What had been noise had gotten so loud it was now impossible to either ignore or explain according to the old model.[28] Counting the number of eddies—the original research question—was one thing. Determining how important these eddies were in the larger ocean circulation was another challenge altogether. The time had come when it was possible to try to determine what exactly was happening beneath the surface. There was a distinct possibility that the Gulf Stream might be less significant, energetically speaking, than the eddies it threw off, an unlikely physical state that would temporarily reverse the otherwise relentless dissipation of energy to ever smaller and smaller scales. It would be the oceanographic equivalent of unscrambling eggs, of a cup of coffee getting warmer rather than colder. The only way to see these eddies more clearly, and to therefore understand them better, was to find a way to measure them more thoroughly.

In the ocean, resolution is a matter of both time and space. The biggest challenge is to observe multiple locations in the ocean simultaneously. This kind of view—called synoptic—was pioneered by meteorologist Robert FitzRoy in the 1850s when he linked coastal observers via a telegraphic network. Some 120 years later, oceanographers were finally on the cusp of doing the same for the ocean. The task was much harder not only because the ocean is a wild and unforgiving place to live and work, but because water is so much denser than air. As a result, it packs much more turbulence into an equal area, and that turbulence—the eddies—lasts much longer than storms in the atmosphere. Ocean eddies are roughly a tenth the size of atmospheric storms, and they last weeks and even months instead of days.

The task that now faced Stommel was to understand how eddies fit into the large-scale structure of the ocean. "We are not interested in describing these eddies in isolation," he wrote in a piece intended to serve as a wake-up call to oceanographers, "we are concerned with discovering whether they play a significant role in driving the large-scale circulation. Is there interaction of eddies and large-scale circulation in the ocean as there is in the atmosphere?"[29] What, for example, is the relationship between these newly discovered underwater storms and the large-scale circulation of the ocean? Do eddies work to dissipate energy, to counterintuitively add energy back into the system, or do they do a bit of both? Stommel had demonstrated that the ocean was susceptible to simple physical explanations. He had given the burgeoning field the confidence to start drawing—sketching—a representation of the movements of the ocean. And he insisted that to know the ocean it must be observed repeatedly, with patience and determination.

For the past twenty years, the way scientists had imagined the atmosphere and the ocean had differed dramatically. The atmosphere had become, thanks to the network of radiosondes, and to the work of Joanne Simpson and others, a turbulent, fast-changing environment. The ocean had remained, in theoretical terms, remarkably still: a "steady smooth flow," in contrast to the "highly nonlinear fluid-dynamical flow, with large eddies [storms] playing an essential and dominant role."[30] It was time, argued Stommel, to determine once and for all whether the ocean was as nonlinear in its motions as the atmosphere. The questions Stommel and Richardson had investigated on the loch were about to be explored on a much bigger scale. "We anticipate that the eddies will be found to play a dominant role in the dynamics of the ocean circulation," predicted Stommel, "and that our whole theoretical concept of ocean currents, developed over the past twenty years, will be changed."[31] It was a risky undertaking that threatened to consign to the dustbin the hard-won theories on which many of the participating scientists had worked, but "if anybody was going to correct or demolish our old theories, we wanted to be the ones who did it."[32]

The best way to determine the answer to this question was not, as Stommel made clear on numerous occasions, to set up a vague and unfocused survey that would passively gather data. "After all," explained Stommel, "anyone can sprinkle dots on a map of the world and call it a plan for future measurement."[33] To really understand the ocean would require a series of experiments with clear hypotheses to test and protocols for assessing their outcome. The planners called the project MODE, for Mid-Ocean Dynamics Experiment. Every word of that tidy acronym was a signal of the reach of its ambitions: to bring motion and the methods of physical experiments to a part of the ocean that had previously evaded observation. The name was a sword to slay the legacy of the descriptive atlases of the kind produced by the *Meteor* that even in the 1970s still haunted oceanography. For this was, quite self-consciously, an experiment—not a survey, not a series of hydrographic stations, but an experiment. The well-defined question this experiment sought to ask was "Do eddies exist on this scale in the deep ocean?" and it set out to do so in a similarly well-defined length of time.

Stommel had identified the "problem" of experiment in oceanography as early as 1963, when he'd written an important paper for *Science*, titled "Varieties of Oceanographic Experience," in a sly reference to William James's classic study on religion. In it, he had made the case for considering every oceanographic expedition as a scientific experiment ("if we regard an expedition as a scientific experiment, then we must propose to answer certain specific questions . . ."),[34] and for taking careful account of the range of scales on which the ocean varies—its varieties of experience. The ocean, it was becoming more and more clear, varied across an astonishing range of time and space. Designing good experiments—experiments that had the possibility of yielding definitive results—meant taking careful account of just what kind of variety existed in the ocean. As the initial results from *Aries* had made clear, it wasn't possible to simply use statistics to average out energy in the ocean across convenient scales. Answering particular oceanographic questions—say the variation in sea level in a particular ocean basin—would require asking questions at the right scale. To this end, Stommel included a diagram which served as a visual index of the range of scales on which energy varied in the ocean, from the hundreds of meter-long gravity waves that lasted just minutes to the tidal variations that happened on a daily and monthly basis, to meteorological effects that transpired on similar scales with much less regularity, to the mammoth variations, occurring over thousands of years and thousands of kilometers, that constituted the ice ages. This diagram was classic Stommel—a deceptively simple tool for ordering complexity. It was both a map of the energy of the ocean and, as Stommel tried to argue, should also be a road map for oceanographers if they ever hoped to catch the most meaningful oceanographic phenomena with the instruments at their disposal. Though the task seemed overwhelming, if matters of scale received their proper due, Stommel thought there was reason to hope that in the future "theory and observation will at last advance together in a more intimately related way."[35]

Working alongside colleagues including Carl Wunsch, Francis Bretherton, and Allan Robinson, Stommel developed a plan to catch an eddy in the ocean.[36] If the holes in the experimental net were too big, an eddy could pass through it undetected. Too small, and they would only see a fraction of it. The entire ocean was, from Stommel's point of view, neither more nor less than a problem in what he called hydrodynamics—the movement of water—on a scale "larger than a laboratory, smaller than a star." Similarly, an eddy was an object of definite size (though what size precisely they could only guess) to which appropriately scaled detectors would need to be directed.[37] Stommel and colleagues determined that a parcel of ocean approximately 300 kilometers square, stretching from the surface to a depth of roughly four kilometers, would be the right size. Like a hunter waiting for its quarry, the scientists would have to set up a trap of detectors and then simply wait, and hope that an eddy would pass in the time allocated to the experiment.

They reckoned it would take six ships, two airplanes, dozens of moorings, neutrally buoyant floats, free-fall velocity profilers, air-current dropped probes, and 121 pressure gauges arrayed on the ocean floor to capture the eddy.[38] To create such a complex system of instrumentation and monitor it during the study would require the coordination of fifty oceanographers from fifteen institutions, including, for the first time, modelers who would be integral to the design and execution of the project.[39] The experiment itself would last four and a half months at a location between Bermuda and Florida. The experiment made use of a new kind of floating instrument—called SOFAR, for SOund Fixing And Ranging floats—that could be sent to specific locations in the deep ocean and moored there, taking measurements of the temperature, speed, and salinity of the water that flowed by them. At the same time, instruments were deployed to follow the currents in an area, floating freely.

If MODE was successful, it could be used to refine the most fundamental theories of the general circulation of the ocean, and would have a corresponding impact on the climate model that combined, or "coupled," the ocean and atmosphere that had been published in 1969 by Syukuro Manabe and Dick Wetherald, at the GFDL at NOAA. Another hoped-for dividend was a better understanding of the ocean—including the possible discovery of the ocean's "weather"—that would improve forecasting for the kinds of weather that affected people on land and at sea. The plan was bold, and the timing seemed auspicious. There was consensus among those whose voices mattered that this was a worthy use of time and money. Nevertheless, there was real uncertainty about whether it would be successful. It was entirely possible that the "eddy trap" they laid would fail to catch anything.

A film made at the time captures the uncertainty (and copious facial hair) of the moment. Scientists sit scattered across a lawn, seemingly performing oceanography in the open air, relaxed, engaged, and egalitarian. "We may all be retired before something's happened," cautioned one scientist, while another worried they might find nothing at all. Deciding where to site the experiment required the scientists to make assumptions about how the oceans operated that the experiment itself was designed to test. There was thus real uncertainty not only about the kind of results that would emerge but about whether anything at all would be measured. A crucial decision concerned whether the experiment should be centered over a relatively flat piece of ocean or a rocky piece of ocean.

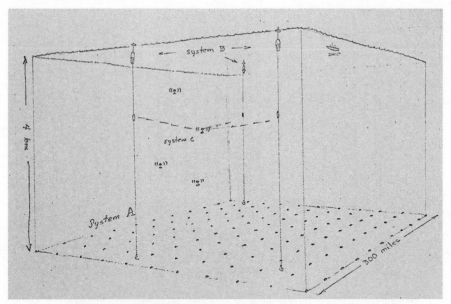

FIG. 6.5. Early sketch of instrumentation for the Mid-Ocean Dynamics Experiment. The experimental area measured 300 miles square and four kilometers deep. On the seafloor, 121 pressure gauges were arrayed with thirty-mile spacing, to be augmented by three to four moored hydrophones and a set of pinging constant-level floats floating at four different depths. These physical systems were to be augmented by a "computer-numerical-model" for prediction of float positions to enable the experiment to be tracked and adjusted in real time. Source: Memo from Henry Stommel, August 11, 1969, Mid-Ocean Dynamics Experiment records, AC 42 Box 2, Massachusetts Institute of Technology Institute Archives and Special Collections, Cambridge, MA. Courtesy of Institute Archives and Special Collections, MIT Libraries.

The answer depended on how much you thought the landscape of the ocean floor—its topography—affected the mixing of the water above it. To what extent could any piece of ocean give the results that oceanographers were hoping for? Carl Wunsch, for one, suggested that "There is no such thing as a typical piece of the ocean, every piece of the ocean is different." When someone countered that the experiment could be centered at the boundary between a rocky and a smooth ocean floor, Wunsch replied that "maybe nobody's going to be satisfied with that compromise." The men laughed, but Wunsch wasn't really joking.[40]

Eventually, the MODE team compromised on a location in the Atlantic (the Pacific was never an option, for reasons that were primarily logistical), choosing a spot that was both rocky and smooth. The experiment itself ran remarkably smoothly, making use of spe-

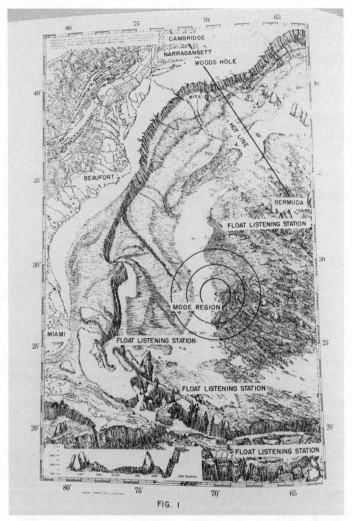

FIG. 6.6. The MODE research zone, straddling an area of smooth and rough ocean floor topography. Source: MODE-1: The Program and the Plan, March 1973, Mid-Ocean Dynamics Experiment, AC 42, Box 1, Massachusetts Institute of Technology Institute Archives and Special Collections, Cambridge, MA. Courtesy of Institute Archives and Special Collections, MIT Libraries.

cial "hot-line" phones that enabled the on-ship scientists to communicate with a headquarters. The only major setback was the disappearance of the central mooring, a hazard of leaving large pieces of equipment unsupervised in the open sea but a puzzling disappearance nevertheless.

The audacious plan paid off. An eddy was found and tracked.

MODE showed that eddies were common and widespread and, most importantly, contained a staggering ninety-nine percent of all the kinetic energy in the ocean. Eddies, in this sense, were the dark matter of the oceans, until MODE revealed that oceanography had been missing the biggest story of all. The mysteries of turbulence and flow were beginning to be fully revealed, if not yet solved—or resolved.

That was a clear success. Less clear-cut was what the legacy of MODE would be for oceanography. MODE had "established a reputation as one of the most tightly run field programs ever, much to the dislike of some of the participants," a writer for *Science* noted. The need for consensus on planning among such a large group meant that, as one scientist put it, "on sticky points, we would take people into the back room and intellectually beat each other into agreement."[41] This was a far cry from the intellectual independence that Stommel had so appreciated. The success of MODE broadcast to the wider world that oceanography could now operate on big scales, with big budgets and dozens, if not hundreds, of researchers working together. It was, as the *Science* article put it, "big science, new technology." But with institutional heft came what many lamented as the loss of individual freedom, and the success of the project was a source of deep ambivalence for many of those who had participated in it. Francis Bretherton, a theoretician who clearly enjoyed the chance to participate directly in a field experiment, nevertheless remarked that "it would be disastrous if the success of oceanographic big science persuades people that it is the only way to do things."[42]

MODE created new ways to think about energy and motion in the ocean. It also, and just as fundamentally, created a whole new way of doing oceanography. Dozens of scientists from several countries had collaborated for an intense and short-lived moment in which an enormous amount of data was generated. For Richardson, for whom the image of a massive calculating organization, consisting of some 64,000 human computers, represented a utopian fantasy, such data would have signaled the arrival of a long-hoped-for future for meteorology. Whether he, an intensely independent-spirited man, would have liked to actually live in such a brave new world is unknowable. Stommel was there to see the day arrive and found himself recoiling from both the administrative pitfalls of big projects and the use of brute force rather than physical theories to crack oceanographic problems. For him, scientific breakthroughs were an intensely personal achievement. "Breaking new ground in science is such a difficult process that it can only be done by an individual mind," he wrote in a *cri de coeur* titled "Why We Are Oceanographers." "For some of us, this is the main attraction of doing scientific work. In this respect, it is like the art of painting or musical composition or poetry . . . it all begins with an individual's choice of medium, choice of theme and style and subject."[43]

Stommel sought to maintain and celebrate the contribution that a single individual (or a group of like-minded individuals choosing freely to work together, as upon an idealized ship expedition) could make to the biggest questions about the ocean. He believed that the biggest questions about the ocean—how it circulated on the basin scale or even the global scale—were accessible only to the best individual minds. But he needed data, not too much or too little, but just enough. To get the data required organization, grant-writing, logistical planning, all of which threatened to consume the time needed for thinking and doing science. The paradox was that the bigger Stommel's ideas about circulation got, the more he needed data to test them, and the bigger the projects got. He could not repeat the 1948 achievement, when he essentially deduced the Gulf Stream from a few equations.

The success of MODE was a source of abiding ambivalence for Stommel. MODE made it possible, even necessary, to think about the old sluggish ocean in an exciting new way. It also raised the possibility that the kind of science that he loved—driven by mental pictures and intense collaboration with a few individuals—would become increasingly difficult to justify. This was a transitional moment, when the discipline stood poised between the old ways (already the object of nostalgia), when it had been possible for individuals to make their own destiny, scientifically speaking, and the future, in which oceanography would be defined by its large projects rather than the ideas of its individual practitioners. This had not yet come to pass, but with MODE, Stommel saw the metamorphosis beginning.

✳

Stommel concluded that the best way—the only way—forward was to "take up various oceanographic phenomena separately as though they were mutually independent (which of course they are strictly not)."[44] To study the machinery meant taking it apart, conceptually, all the while remembering that the machine only functions when it is whole. The questions raised by MODE, in other words, could only be answered with other experiments similar to it which could probe different elements of the ocean system. And so, despite his deep, almost constitutional, misgivings about the new "bigness" of oceanography, Stommel continued to involve himself in projects similar in scale to MODE, including a follow-on project, a contentious collaboration with the Russians, called POLYMODE. The pace with which these new experiments occurred was intense. Between 1973 and 1978, a total of nine field experiments were planned, an entire alphabet soup of acronyms from MODE to GARP, NORPAX, JASIN, CUEA, SDO, INDEX, ISOS, and GEOSECS.

Thanks to these experiments, it became clear through the course of the 1970s that the answer to the question of where eddies existed was relatively straightforward. Almost everywhere people looked, they found eddies.[45] Eddies in the North Pacific. Eddies in the Arctic. Eddies in the Indian Ocean. Even eddies in the Antarctic. The question become not where eddies were, but where they weren't. By 1976, John Swallow was wondering whether there was any place in the ocean from which they were absent. It appeared there was not, and this fact, added to the energy that they contained, made it conceivable that eddies were not incidental to ocean circulation but essential to it.[46]

Near the Gulf Stream, rather than decreasing the energy of the Gulf Stream, eddies seemed to be adding to it. In other words, the viscosity of the eddies—that is, the extent to which they acted as a brake or drain on the energy of the system—was found to be negative. Looking at the data, a theoretical oceanographer named Peter Rhines wrote that it was possible to see that "the small eddies have coalesced to form a few lazy gyres. This is just the reverse of the result for totally chaotic, three-dimensional turbulence in which energy is degraded into smaller and smaller eddies until ultimately being lost to viscous smoothing." Given this surprising and counter-intuitive coalescence of eddies, Rhines noted that it would be neces-sary to rethink our understanding of how turbulence functioned in the ocean. The classic example used to explain turbulence—a cup of tea into which milk has been stirred—was no longer a reliable model. "Clearly the ocean is not a teacup," concluded Rhines, "and energy put into eddies and intense currents cannot simply flow out of the system into minuscule eddies and thence be dissipated by viscosity."[47]

If eddies were *adding* energy to the system rather than enabling it to dissipate, then any description of the circulation that simply ignored them (because they were too small to see) would be fun-damentally inadequate. Whatever theories existed for explaining the large-scale circulation were suddenly much less secure. Joanne Simpson and other meteorologists had already covered this ground (so to speak), in the air. It had taken roughly fifty years, beginning at the turn of the century, for them to realize that storms in the atmo-sphere were not merely a means by which the system threw off excess energy. Instead, they came to understand that storms—what we would call "weather"—in fact feed energy back into the system and affect climate on the largest of scales. Rhines imagined what the implications of negative viscosity in the ocean would be for a world circulation. "Eddies, if fast enough, can gang up to drive a system-atic flow. It is just possible that the ocean works as a sort of Rube Goldberg device, with the wind driving a strong circulation that breaks down into eddies; the eddies then drift off and radiate into the far reaches of the ocean, where they recombine to drive new ele-ments of circulation. Models of this sort are now being explored."[48]

Personally, Stommel worked hard to reconcile the need for col-laboration in order to answer the "big and difficult" problems posed by the ocean with his desire to keep oceanography a bureaucracy-free zone. He characteristically found inspiration in the ocean itself.

FIG. 6.7. Henry Stommel in conversation with George Veronis. Photo by Vicky Cullen. © Woods Hole
Oceanographic Institution.

"We are beginning to see spontaneous and fluid groupings of oceanographers," he declared hopefully, in a document written for just
the sort of government program he might have wished to steer clear
of, "whose aim is to grapple with certain long-period and large-
scale phenomena in the ocean." These eddies of researchers come
together for time-limited events—called "experiments"—which "the
scientists involved expect to carry out themselves."[49] Little auton-
omy would be lost if these groupings could somehow be relied upon
to spontaneously form and dissipate. Though MODE was "a form
of Big Science," he hoped it could be just a temporary form, summoned into being for a "particular job and dissolved in a few years
when that job was done."[50] The contrast with studies that, as Stommel put it, "depend on large volumes of data gathered in routine
fashion by governmental agencies" could not have been stronger.
If Stommel was polite enough to acknowledge that both types of
investigation were useful, it was more than clear on which side his
sympathies (and creative energies) lay.

All of this research took a toll on Stommel, who was central to the planning of almost all of it and the implementation of a significant portion of it. Stommel, who was accustomed to publishing up to six papers a year (often with coauthors), published nothing at all between 1974 and 1976. It didn't help that he had been living since 1960 in a self-imposed exile from WHOI, first at Harvard and then for fifteen years at MIT. He had left WHOI in the year that Paul Fye had taken over the directorship of the Institute, having found it impossible to work under him. With Fye in charge, the wolves of bureaucracy seemed to be circling ever closer. This period of creative drought only ended when Fye retired, and in 1978, after eighteen years, Stommel finally felt he could return to his intellectual home at Woods Hole. Then, in his words, "I gave up teaching and all administration and began to live again."[51]

All of these anxieties about both the changes in how oceanography was practiced and the possibility of understanding the ocean globally were to a certain extent internal to the field of oceanography, reflecting its particular disciplinary history and the coming-of-age of the postwar ambition to create a physical science of the ocean. By the early 1970s, in oceanography as in meteorology, new external pressures were beginning to play a decisive role. The military applications of oceanography that had garnered it funding and prestige during and after the war were, by the 1970s, increasingly being supplanted by the climate-predicting utility of oceanography. By 1974, there was a growing awareness that it would not be possible to understand changes to the earth's atmosphere as a result of rising carbon dioxide without understanding the ocean as well.[52] It was incumbent upon oceanographers to determine how "to make progress toward building their oceanic part of the model," according to a National Research Council panel on "The Role of the Ocean in Predicting Climate."[53] A steering committee, chaired by Stommel, was set up to investigate the "large-scale ocean-atmosphere coupling (particularly as they relate to the ocean's effect on climate)."[54] The models to which Rhines had referred in his report on the role of eddies in the general ocean circulation were numerical computer models, which had, by the 1970s, assumed a central role in the way climate variability was investigated. At the end of the decade, the National Research Council appointed a committee, which came to be known as the Charney Committee, to consider the role of rising carbon dioxide and major climate change. Stommel was one of three oceanographers in the group. Their report made some guesses about future global mean temperatures and noted strongly that much remained unknown about the ocean's response to atmospheric warming, its uptake of carbon, and its thermal memory.[55]

The drive to consider the ocean as part of a global climate system coincided with the possibility of seeing the earth from space. Such a prospect had been seriously considered ever since 1960, when the TIROS series of weather satellites had returned the first images of cloud patterns taken from space. Oceanographers realized that if satellites could one day provide them with sufficiently accurate measurements of sea surface elevations (to within around fifty centimeters), it would be possible to map the location of ocean currents from space because these currents are warmer than surrounding waters and cause the ocean to bulge upward. Fifteen years later, what had once seemed like science fiction became a reality. In 1975, *Geos* 3 provided the first comprehensive picture of the ocean geoid, the bumpy average sea level that would be visible if the effects of tides and waves could be magically subtracted from ocean surfaces, leaving only the variable effects of gravity on the ocean. To prove that the data from *Geos* 3 was good, scientists used it to find a so-called "cold core ring" or eddy, which had been compared with data collected from buoys in the sea, aircraft in the skies, and infrared (versus altimetrical) readings from satellites.[56] In 1978, the picture from space got sharper still when SEASAT provided even more precise sea surface readings, revealing the presence of the Gulf Stream.[57]

In addition to the new global visions of the oceans that satellites seemed on the cusp of being able to deliver, there was another global ocean that seemed well within reach. This was the global ocean created by the climate modeling community, a fast-growing group of researchers who relied upon advances in computing power to enable them to calculate ever more finely resolved numerical models of the earth, in which the grid which was laid down, numerically, over the earth, became ever tighter. These models held the promise for a new kind of global knowledge in which physical equations mimicked the actions of real water. Without data from the actual ocean with which to check—or calibrate—these models, however, they risked becoming elaborate fictions with no connection to reality. Unlike the velveteen rabbit, the models demanded not love but data to make them real.[58]

The same year that Stommel helped chair the NRC panel on ocean-atmosphere coupling, a meeting was held in Miami by a new group called the Committee for Climate Change and the Ocean. The question of the ocean's role in climate had become increasingly pressing, thanks to the work of the modelers who demanded data to calibrate their tools and to the growing understanding of scientists working in a range of climate-related disciplines that the earth's climate was a global and coupled system, of which the ocean was a critical component. Building on this growing awareness, oceanographers and other researchers met in Miami to consider climate change and the oceans. There, Carl Wunsch suggested that in order to understand the ocean's contribution to the climate, it would be a good idea to at least try to measure the ocean circulation globally.[59] And so the stage was set for a new project that brought together two important groups of researchers, whose fates would henceforth be intertwined.[60] The project brought together physical oceanographers studying the circulation of the ocean as a problem in ocean dynamics alongside newly christened climate scientists (inheritors to some extent of climatologists' concern with average temperatures), who wanted to understand the relationship between the ocean and the atmosphere as it related to the uptake of man-made carbon dioxide.[61]

Like MODE, the new project was also an experiment—the World Ocean Circulation Experiment (WOCE). Also like MODE, it was a program designed to answer a particular question. It just so happened that the question was a big one: What is the nature of the world ocean circulation? The breadth of the question begged another one: Was WOCE really still an experiment, in the sense of a time- and process-focused singular event? Or, to get anywhere close to an answer about the world ocean, would it be necessary to set up a program on such a scale that WOCE would become a newfangled version of the old-style, observation-rich, theory-poor hydrographic surveys?

Stommel stayed well clear of WOCE. Though he had always been attuned to the ways in which theoreticians, modelers, and observers related to each other, and emphasized the need for close contact between them, the scale of WOCE was too big for him. With MODE, he felt, the size of the project had worked well. But when it came to larger undertakings such as WOCE, the connections between researchers, he feared, became necessarily more rigid and more problematic. The bigger the model, the more observations it required, and the more organization. The way MODE had functioned—"a loose association of individual investigators to raise the neighbor's barn" is how Stommel put it—would not be sustainable over the time frames needed to support the longer-scale models. "If you think about joining one of these longer-term, big-scale things, you're really sort of becoming an employee of a construction company. And that's not very attractive to many of us. We're a little bit scared of the long-term commitments that might be involved in programs like this."[62]

It took roughly thirty years from the first planning stages until the data from WOCE had been fully analyzed. In the process, WOCE helped transform oceanography. In 1985, it had been hoped that WOCE would "provide the first comprehensive global perspective of the ocean as an element in the planetary climate system . . . the driving force for attempting a WOCE is the recognition that predicting decadal climate change will depend on accurate calculation of changes in the large scale flow of heat, freshwater and chemicals in the oceans."[63] It had achieved this, enabling a simplified but powerful quantitative estimate of the ocean's role in transporting heat, which could be combined with that of the atmosphere to offer a model of the earth's climate at the global scale.

As impressive as this achievement was, perhaps the greatest legacy of WOCE was to more meaningfully reveal the ignorance of oceanographers and what shape a program meant to ameliorate that ignorance might take. Pondering the future of their discipline in the early 1980s, while planning for WOCE was ongoing, Carl Wunsch and Walter Munk had written that "we can now appreciate the magnitude of the job facing oceanographers who wish to understand how the ocean 'works,' and who might one day hope to forecast changes in ocean conditions. The ocean is a global fluid, not unlike the atmosphere, and one wishes to observe the global system on all important space and time scales."[64] Envisioned as a snapshot of the ocean of unprecedented scope, WOCE engendered the felt need for a sustained program of global observation. It wasn't a return to the old survey days but it was, in its own way, a return to the discipline of sustained looking. The difference was that as a result of rising concerns about climate variability, looking had become monitoring. The oceans could no longer be simply a place to search for knowledge (if they had ever been that). They had become a signal of a changing global climate that had to be monitored.

Stommel had hoped that WOCE would be the best of both worlds, a blend of the traditional and novel, between the comprehensive and the specific: a compromise between geographically oriented surveys and process-oriented experiments designed to test key physical hypotheses. "Perhaps," he wrote in a heartfelt 1989 essay, WOCE "isn't really Big Science at all—just an assemblage of the miscellaneous smaller projects that people would have wanted to do anyway."[65] This sounds like a man trying very hard to convince himself of something. He sounded more convincing when he considered the unknown scientists yet to come, young people who would challenge the accepted notions of his generation, who would risk their own predictions and who might once again remake the oceans in a new image. Whatever form science might take in the future, whatever organizations were deemed necessary to answer the questions scientists seek to pose, Stommel still believed that for any one individual who seeks to do science, it was the personal endeavor, the "wrestling match with some aspect of the universe," that was the main challenge and central reward. "All alone, one confronts the unknown and divines some meaning from it. We sort the pieces and arrange them in new patterns." In the ocean, as Stommel knew better than anyone, the pieces are innumerable, the patterns more infinite still.[66]

Given the depth of his feeling for what the personal experience of oceanography could be, it is more than a little jarring to listen to Stommel, in a recording made in 1989, sharing his thoughts on the relationship between science and faith. Rather than providing answers to questions that had previously been unthinkable, Stommel here describes science as limited in its scope. "It seems to me that science," he says, "is really very restricted in what it can tell us about the world, how it can meet our needs, the things that we desperately want to know." Just three years before his own death at the age of seventy-two, Stommel describes a visit to the deathbed of his friend and mentor Ray Montgomery (the man who had suggested to him back in 1947 that it might be worth considering why streamlines crowd on the west). Stommel asked Montgomery his thoughts on the meaning and wonder of life and what he thought it all added up to. Montgomery replied that his mind was closed on the subject. He offered neither Stommel nor himself any solace in the face of death, remaining steadfast in his refusal to seek the comforts of spirituality. The exchange made a deep and disquieting impression on Stommel.

If he did not himself reject spiritual values, Stommel was rigorous in separating them from his scientific activity. Though he had spent his life wrestling with—and finding great joy in—the mysteries of the universe as manifest in the ocean, at the end of his life he was adamant that it was dishonest to arrogate to science the kind of wonder long associated with religion. "We have ideas of reverence," he said, referring to the way people sometimes thought of science as a kind of religion. "We get a great thrill out of going into Westminster Abbey and seeing Newton's grave and Kelvin's grave. I regard the library here [at Woods Hole] as some kind of a temple. Now what in the world do words like beauty and reverence and temple have to do with science as we know it?" he asked. "I don't have an answer to that," he replied, pausing before adding, "This is a subject which troubles me a good deal."

By the end of his life, it seems that Stommel's own desire for wonder had far outstripped his sense that science had any claim to provide it. There are echoes here of Tyndall's compulsive revisiting of the idea that science produced wondrous things—including the human capacity for wonder—for no reason at all. But while Tyndall never failed to be satisfied with the wonder that his appreciation of nature produced, in Stommel's remarks there is a sense that his frustration at the limits of science has outstripped his awe at the productions of nature.

Stommel was in agreement with Tyndall on this materialist point. Science was, he clarified, "like the instructions on a microwave oven." It was "awfully dry and dead and unkind," devoid of any of the moral and emotional value that gave human life meaning. The problem was not that science couldn't deliver these things, but that we made the mistake of expecting it to. Science dazzles us with its successes, Stommel explained, "and then all of our hopes and wishes and anxieties somehow diffuse into it and then we talk about the beauty of science and our love of it and our reverence for it, and there isn't anything like that in it at all as far as I can see."

He hoped that the new limits to what science could hope to know—limits that he himself had helped uncover in the turbulence that lay within the very machinery of the ocean—might help recalibrate the place of science in our emotional landscape. Maybe, in other words, we could learn with time to expect not more but less from science, and to be better off as a result. "My own feeling about science is that for me it's been useful as a diversion from more important things. It has been a measure of relief from getting crushed by wonder and anxiety. It's a form of whistling in the dark."[67]

7

OLD ICE

One rainy Saturday in June 1952, Willi Dansgaard stood in his back garden in central Copenhagen. He set a beer bottle on the ground and put a funnel in its mouth. The rain was thick and persistent, the product of an epic storm front that stretched all the way to Wales, one thousand kilometers away.[1] In a few hours, he stepped again into the back garden. The bottle was nearly full. He poured its contents into a small container, sealed the container, and replaced the empty bottle on the back lawn, like any amateur meteorologist collecting rainwater. He re-entered the house and wrote the date and time on his sample, setting it next to the others he'd already taken. The containers crowded his kitchen table.

He kept it up throughout the weekend, even waking in the night to collect the rainwater. The rain, making up an unusually well-developed warm front followed by a sharply delineated cold front, continued. At some point, he ran out of sealed containers and started collecting water in pitchers and pots from the kitchen. He kept it up through the early hours on Monday morning and then loaded the whole collection into his vehicle and drove carefully to his laboratory.

A mass spectrometer, the instrument that made the collection of all this rainwater something more than an exercise in amateur weather-watching, was waiting for him. It had been his since he arrived back in the city after a few early years of adventure in Greenland. He'd fallen in love with the harsh beauty of the vast island, but he had been offered a salaried research post at the university in Copenhagen that he couldn't refuse. The task he'd been assigned was to use the chamber of the mass spectrometer to test the utility of the so-called stable isotopes for medical or biological uses. Radium and its highly unstable cousins had been used for nearly fifty years to treat cancer, helping burn away sick cells or track the progress of disease, but they were damaging to the healthy tissues of the body too. Dansgaard's job was to investigate whether something useful could be done with the less protean isotopes of oxygen and nitrogen. These non-radioactive elements were free of the dangers that accompanied radiation treatment.

Dansgaard had little luck researching the use of stable isotopes in medical applications, but the machine was his to use more or less as he pleased. He knew that the oxygen in rainwater existed in the form of several isotopes, variants distinguished by the number of neutrons in their nucleus. The powerhouse ^{16}O, the isotopic bully, was vastly more prevalent than the heavier ^{17}O and ^{18}O versions. Generated in the belly of stars, at the tail end of the process of helium fusion, ^{16}O makes up more than 99.7 percent of all the oxygen molecules in the earth's atmosphere. That leaves about one molecule of ^{18}O for every three hundred molecules of ^{16}O. The difference of a few neutrons between these isotopes doesn't matter chemically. But it does mean that ^{18}O, which has two more neutrons than ^{16}O, weighs ever so slightly more.

On this basis, the isotopes of oxygen could be separated in the mass spectrometer. He'd done as much. With the machine, it was possible to tease apart the scarcely before separated strands of oxygen with relative ease. He knew that water evaporates in warm temperatures and condenses in cooler temperatures. One form of this condensation is rain. He also knew that ^{18}O, being heavier, was about ten percent more likely to condense than ^{16}O. The converse was also true: ^{16}O, being lighter, was ten percent more likely to evaporate than ^{18}O. Storms were made up of weather fronts, alternating masses of warm and cold air. Putting the two ideas together, Dansgaard wondered whether rainwater had a distinctive isotopic composition, like a fingerprint. Could it change, from rainfall to rainfall, or even during the course of a single storm?

The bottles sloshing in his car were snapshots across time of the isotopic profile of the storm that had just dumped so many inches of rain on Copenhagen. He put the samples through his mass spectrometer, and the results were more than good. It was as though he'd put a stethoscope to the storm and listened to the heartbeat of isotopic oxygen pulsing within it. The proportion of ^{18}O in the rainwater he collected rose as the warm front passed over Copenhagen, as the "heavy" water condensed as rain.[2]

It was all there waiting to be measured and understood, a secret code to the life history of the storm, its changing temperature profile captured in an isotopic progression. He knew immediately that he now had to get inside a cloud. That was the next step, to understand the relationship between the atmosphere at the center of a cloud, that which lay at its top edge and that which was below it. He imagined cumulus clouds as giant condensers in the sky, separating water vapor, sending warm air upward, cold ^{18}O-dense water below, and finally warm ^{16}O-dense air to evaporate at the top.

A buddy with connections to the Royal Danish Air Force got him and his wife, Inge, in a plane. Inge insisted on going along for the ride. She did not, she said, wish to be a young widow. So up they went into the violent air currents inside a cumulus cloud. Using cold traps cooled by dry ice, Willi tried to collect cloud droplets as they tossed about. The samples were too small to quantify the effect, but the results confirmed his theory. Dansgaard looked for a way to gather larger samples, samples that would speak more loudly. He thought of weather in three dimensions: not simply what fell to earth, or what the thermometers measured at a set of points, but masses of gas and liquids traveling together, clouds occupying space the way an actor fills the stage. Clouds had form, depth, density. They moved and changed. And they were freckled with the stable isotopes of oxygen.

The whole water cycle was his to explore. Where would he go next? His mind flew to the rivers, which were liquid averages of the rain that fell across vast distances. He was leaving behind the

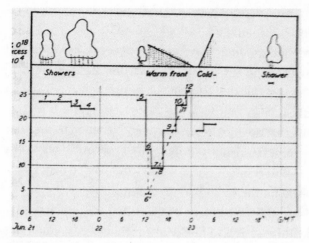

FIG. 7.1. A schematic diagram showing the excess ^{18}O in rainwater samples collected during the June 21–23, 1952 storm. During the passage of the warm front, the rain became colder and isotopically heavier. This was the beginning of Dansgaard's idea for an isotopic thermometer. From Willi Dansgaard, "The Abundance of ^{18}O in Atmospheric Water and Water Vapour," *Tellus* 5 (1953): 461–469. Copyright the Author 1953.

temperature profile of a single storm to think, instead, about what weather adds up to when it is averaged: to climate. Would rivers bear witness to the regions from which their water was gathered, presenting an isotopically determinable average temperature for the region? His mind raced outward, searching for a way to grasp the whole earth with his new insight, his new tool. What he needed was a global network for rainfall collection. And that, thanks to the offices (literally) of a Danish shipping company with international reach, is what he got. Another buddy with connections managed to put him in good graces with the Danish East Asiatic Company, which obligingly provided him with samples from its worldwide branches. Soon enough, he had a new collection of bottles, this time sloshing not with Copenhagen rain but with the products of storms ranging from the tropics to the Arctic.

He was now equipped to continue his stepwise progression through the waters of the globe. He was working outward and upward. Would the relationship between the isotopic ratios and temperature that he had observed in his backyard samples hold true for water from other parts of the globe? At first, it didn't look good.

Things got messy in the tropics, and it was hard to discern a strong correlation between the δ-value (a measure of the isotopic ratios) and the temperature of the water. But all was not lost. Samples collected closer to the poles, and, in particular, at high altitudes in the Arctic, showed the same relationship he'd been so happy to find in the Danish rain. Thanks to the Danish East Asiatic Company, he had proved that the isotopic technique could in principle be used to measure temperature at the time that rain or snow fell.

It was a good result, but Dansgaard was not satisfied. In the paper he published on the subject in 1954, Dansgaard wondered whether the correlation between isotopic ratios and temperatures in cold places like the Arctic also held true for the distant past of the earth. "On the supposition that the character of the circulatory processes, in all essentials, have not varied over a long period of time," he reasoned, the isotopic technique could offer the possibility of "determining climatic changes over a period of time of several hundred years in the past."[3] This was the first inkling of what he later called "maybe the only really good idea" he ever had.[4] Maybe, thought Dansgaard, he could use his mass spectrometer to measure not only today's rain, but yesterday's snow, and many, many yesterdays before that. He thought he might be able to go back in time a few hundred years because the only way to access ice at depth was to measure it at the margins of the glaciers, where it was tens of meters thick. He ended his paper with the statement: "an investigation will be undertaken as soon as an opportunity offers."

Dansgaard's idea depended on new technology and a newly developed understanding of atomic isotopes, but the problem to which it would eventually contribute with stunning results was of a piece with old questions about the role of ice in the earth's history. Thanks to the work of men like Agassiz, Forbes, and Tyndall, the ice age theory gradually grew in acceptance through the second half of the nineteenth century. Much of this had to do not with the convincing power of the theory itself, but with the accumulation of what amounted to an avalanche of evidence in its favor. Mapping, in par-

FIG. 7.2. Map of Europe showing largest extent of glaciation, from William Bourke Wright, *The Quaternary Ice Age* (London: Macmillan, 1937).

ticular, played a critical role in turning the ice age theory into an ice age consensus by bringing together countless observations of glaciers, ice sheets, and the moraines they left behind. By the 1870s, enough evidence had been plotted on maps to convince the most skeptical that erratic and uncontrolled drifting icebergs could no longer be considered a plausible mechanism for the degree and nature of the evidence compiled.[5] What James Geikie called the "prejudices" of British geologists in favor of icebergs (in a letter to John Muir) had been overcome by the sheer mass of evidence in favor of ice sheets. In Britain, the culmination of this mapping mode arrived in 1914 with the publication of William Bourke Wright's comprehensive *The Quaternary Ice Age*, which indicated the flow lines of glaciers believed to have moved across Britain, re-creating the movement of the ice based on the moraines and striations it had left behind.

Maps like Wright's helped convince the remaining skeptical geologists that not just one but several ice ages had in fact occurred. "There is a growing body of evidence that several distinct glaciations were involved," explained Wright.[6] But it still remained unclear precisely what had caused the earth's climate to flip-flop between ice

FIG. 7.3. Two views of a glaciated boulder showing the different directions of the striations. From William Bourke Wright, *The Quaternary Ice Age* (London: Macmillan, 1937).

ages. James Croll had made many assumptions about the complicated interactions between different aspects of the earth's climate system, with ice at its poles, water vapor that created cooling fogs or reflective clouds, and ocean currents that meandered across the globe transporting heat all playing a role. While Croll and his supporters were convinced that the laws of physics meant that such a system had to exist, the idea that the oceans, ice, and atmosphere interacted at global scales was both new and almost impossible to prove. Geologists who were used to basing their grand theories on extensive field observations and synthesizing maps were unconvinced. And so it was that Croll's theory remained marginal to most discussions of ice age theory in the decades following his death.

One man took up Croll's astronomically based theory with an almost revolutionary zeal. Milutin Milankovitch was a Serbian engineer who ended up as a prisoner of war during World War I. He used his four years of imprisonment to begin to work out detailed calculations of the changing orientation of the earth and the sun over hundreds of thousands of years. Three main cycles, corresponding to the tilt and wobble of the earth on its axis and slight changes in the shape of its orbit around the sun, which changed over long time cycles ranging from 23,000 to 41,000 to 100,000 years, determined how the sun's radiation hit the earth. While they did not alter the total amount of sunlight that reached the earth, they did affect the distribution of that sunlight throughout the year and across the globe in ways that affected the earth's climate dramatically (as Croll had understood). Interestingly, Milankovitch shared Croll's belief that changes in the earth's orbit were the cause of the ice ages, but he thought that increased summer sunlight in the northern latitudes where snow and ice accumulate and then melt each year, rather than increased winter cold, was the main reason for the imbalance.

Like Croll, the Serbian struggled to elicit mainstream recognition for his work. But also like Croll, he received support from a few eminent scientists. One important group of researchers called themselves climatologists. Concerned with mapping as well, these scientists sought to establish not the past history of the earth but the current geographical distribution of different climates. Once content to focus on the local or regional scale, climatologists had by the turn of the century become increasingly confident in their ability to map climate differences around the globe. Place, rather than time, was their main concern. These intellectual descendants of Alexander von Humboldt were not concerned, as the geologists were, to explain change over time. More attuned to the stability of regional climates than to any long-term changes, they were inclined to stay focused on the climates of the present. Cleveland Abbe, chief scientist of the U.S. Army Signal Office, for example, considered that "the true problem for the climatologist to settle during the present century is not whether the climate has lately changed, but where our present climate is, what its well-defined features are, and how they can be most clearly expressed in numbers."[7]

The concerns of climatologists were shared by—and partly a response to—nations seeking to consolidate or increase their territory. Mapping had always been a tool of government. Where to plant which crops, how many settlers a region could reasonably accommodate, how much rainfall fell in which regions were climatological matters of central importance to state powers. Since climatological regions did not always follow national boundaries, one nation's drive to map its territory naturally spilled over into others—prompting international cooperation even as individual nations sought to better manage their own territory. Climatology gradually grew in both prestige and ambition. By the early decades of the twentieth century, climatologists were increasingly determined not merely to characterize distinct regional climates but to create a global climatology that could synthesize local information. Given the new global orientation of the field, it makes sense that Milankovitch's work on global climate changes would catch the eye of perhaps the most accomplished climatologist alive, Wladimir Köppen. Köppen, too, took a global view of climate. In 1884 he produced one of the first maps of climate on a global scale, plotting regions of similar temperature and precipitation, flora and fauna across the globe. Köppen and his son-in-law, the meteorologist and Arctic explorer Alfred Wegener, were both impressed with Milankovitch's work.

Soon after his first paper came out in 1920, Milankovitch received a postcard from Köppen praising his publication, which he cherished, he later said, "like a relic." The product of Milankovitch's decades of toil was the first time line showing how summer sunlight had varied over the past 600,000 years. While Croll had made qualitative guesses about how radiation might have changed, Milankovitch had managed to generate a curve giving specific dates for astronomical oscillations that would have affected how much sunlight hit the earth during the summer. Köppen and Wegener realized that with Milankovitch's curve, they had an independent marker of past climate change. Geologists had already generated their own crude time line of past ice ages based exclusively on geological traces—on the moraines, striations, and other features that had first been noticed back in the 1830s and 1840s. There was rough agreement between these curves when they were aligned, giving both geologists and physicists more confidence that changes on the earth might correspond to astronomical cycles. It was now possible to make retrospective predictions about when ice ages and warmer periods may have occurred, and to do so on the basis of a theory with an impeccable physical basis. The long arm of Newton stretched across the twentieth century, demonstrating that something as seemingly contingent as the spread of snow and ice across the earth was intimately related to the celestial mechanics of the earth and the sun.

It took some time for geologists to come around to Milankovitch's theory. During the 1930s, as he refined and extended his calculations, geologists sought to match the curves of past ice ages generated by increasingly detailed geological evidence to the ever more detailed astronomical curves that Milankovitch produced. The concordance was convincing, even if there were significant divergences between the curves as well. Over time, something important shifted in the relationship between Milankovitch's theory and the field-based work of the geologists. No longer was the geological record being used to test Milankovitch's theory. Now the theory was used to verify—or ground-truth—the geological record. "In this manner," wrote a triumphant Milankovitch, "the ice age was given a calendar."[8]

Milankovitch had reached into the heavens to produce a calendar for changes on the earth. His theory leveraged the predictive power of astronomy to make inferences about the physics of the earth. In a sense, this was the culmination of ambitions laid out by men such as John Herschel and Norman Lockyer, the "cosmical physicists" who hoped to simultaneously draw the connections between Earth and the solar system and to raise the physics of terrestrial phenomena to the same status that positional astronomy had once held. Milankovitch's reliance on Newtonian calculations of orbits rather than on the physical traces of magnetism or radiation marked him out as different from Lockyer and Herschel. And while Milankovitch's theory had been corroborated by geological evidence on earth, it was still a rough-grained picture of the history of the ice ages. It was possible, and indeed likely, that other changes relating to the so-called secondary factors which Croll had initially identified— factors such as the circulation of the world's oceans and changes in atmospheric phenomena—occurred on much shorter timescales than the cycles Milankovitch had identified. To find out what these were would require new tools for looking into the past. The heavens had provided what they could. The time had come to look back to the earth and to seek new methods, which went beyond the descriptive mapping into which the geologists had put so much effort, to bring the past into focus.

The answer to the implicit question raised by Milankovitch's work—What traces would such cycles have left on the earth?—led to a new field. This field, called paleoclimatology, combined the descriptive elements of climatology with new physical tools to generate a map (or series of maps) of the earth's past climates. The location and nature of rocks had long served as crude proxies for temperature. Using the geological traces—the literal scratches left by glaciers as the ice passed over bedrock, or the moraines deposited when they melted—it was possible to infer that temperatures had been low enough to support ice or warm enough to melt ice. Any finer degree of resolution was impossible. To see with more acuity, it would henceforth be necessary to make a literal rather than figurative journey into the earth's past, to drill quite literally down into the planet to extract from ice sheets the cold remnants and from muddy ocean sediments the undisturbed detritus of deep time. These cold and muddy archives could only be read in the 1950s, thanks to new tools that relied on the same physical understanding that had built the atomic bomb. Paleoclimatology was built largely (though not entirely) on the foundations of nuclear physics and came to fruition in the postwar world of anxiety and optimism engendered by the power of the bomb.

Having come up with his idea for an isotopic thermometer for the past, Dansgaard was gripped by the simultaneous urge to share his idea and to protect it. He wanted priority, but in order to stake his claim, he also had to publish it, or at least enough of it to make clear what it was. The problem was that he hadn't yet proved it would work. So he wrote up his paper in which he had suggested vaguely that it might be possible to use his technique to peer into the past "several hundred years" by analyzing Greenland glaciers.[9] He mentioned Greenland because of the wondrous fact that the snow there never melted in the summertime. A record of each year's snow was preserved under the next year's snowfall, creating an ice sandwich several kilometers thick in which every layer represented a distinct year. Though he did not know it at the time, this record stretched back not hundreds, or even thousands, of years, but tens of thousands of years.

Dansgaard set off to find some old ice. He figured his best bet would be to catch icebergs as they calved from glaciers into the sea. He started in Norway, with a charismatic Norwegian named Pete (Per) Scholander who believed that air bubbles trapped in the ice were tiny atmospheric time machines, holding samples of the air from the time when the snow fell around them. Not only could the bubbles be made to speak of their contents—what mix of atmospheric elements such as nitrogen and oxygen was present in the past—but, thanks to the new technique of dating the CO_2 content in the bubbles using the carbon-14 isotope, they could tell the age of the atmosphere they contained. It was a perfect fit for Dansgaard's interest in using oxygen isotopes to study past climates. Together, Dansgaard and Scholander set out for the Jotunheimen massif. Within six weeks they'd managed to pry two very large pieces, each weighing five tons, from the old and young ends of the glacier. They melted the ice and captured the released air. Carbon-14 dating gave its age as 700 years old, in agreement with estimates by glaciologists. But there was an unforeseen problem with the relatively young snow of the Norwegian glacier. As the water melted, it took with it some of the most soluble gases, skewing the ratios of the gases in the atmospheric air. They needed colder ice, untouched by meltwater, to avoid this problem.

And so one year later, Dansgaard finally found himself heading back to Greenland, on a much more involved expedition with Scholander, which he dubbed the Bubble Expedition of 1958. They sailed around Cape Farewell at the southernmost tip of Greenland to the western coast. There they set anchor and erected a bulky laboratory on the front of their boat (to the dismay of the captain). They waited for the giant glaciers on the coast to calve into the sea and then set about spearing chunks as big as they dared, harpooning the great white blocks as if they were whales. Or they rammed their ship into smaller icebergs and hoped for pieces of the right size to result—holding on to fragile glassware in the shipboard laboratory as they did so. It was a scientific adventure, full of improvisation and high spirits.

They melted ice day and night, and their "bubble" harvest from a summer spent herding icebergs was eleven samples of carbon dioxide coaxed from the bubbles in the heart of the glacial ice, each a distillation of some six to fifteen tons of ice. The science, and the effort, seemed designed to reduce maxima of ice and effort into minima of carbon dioxide and data. This suited the men, who were seeking an elusive signal from the past. Clearing away masses of extraneous material was just what was needed to get to the signal they wanted to capture.

Dansgaard himself returned with not dozens but thousands of plastic bottles, each filled with a melted ice sample, a library, or frozen annal as he called it, of the earth's past climate. He put his mass spectrometer to work on them and soon had a mini-production line going, producing twenty isotopic ratios, or delta values, a day.

Ice looks different when it has been flooded with meltwater, and Dansgaard could tell just by looking at the sample of the glacier ice in an ice chunk they studied which layers had meltwater and which didn't. His delta values matched up to these visible layers. It was another good sign for his good idea, but the ice they'd sampled wasn't old enough to prove what Dansgaard really wanted to know: Could temperatures be read off the oldest ice in Greenland? Answering the question would require getting off the water and into the interior of Greenland, where the ice sheet was deepest. Hundreds of meters beneath the surface of the ice lay the records of the past climate. Dansgaard was convinced of it.

As Dansgaard raced about trying to collect samples from rivers, storm clouds, icebergs, and glaciers around the world, his mass spectrometer machine sat in Copenhagen, the ultimate destination of the water that Dansgaard collected and the reason it was even worth doing so. The machine did not, in fact, analyze the water samples themselves. It was easier to transfer the oxygen molecules from the water into a bit of gas, carbon dioxide, and then analyze that by causing the molecules to diverge according to weight.

The water that Dansgaard collected contained more than just a ratio of isotopic variants. When he stood out in his back garden collecting rainwater in 1952, he was also collecting traces of radioactive elements released into the environment—the atmosphere, the ocean, and the earth itself—by the explosion of atomic weapons. Aside from the two bombs dropped on Japan by the United States, these weapons had been exploded not to wage war—at least not directly—but to prepare for the possibility of war by testing the effects of different types of weapons.

Both the bombs and the mass spectrometer Dansgaard used to trace the isotopes were evidence of the success of a new kind of physics, the physics of the nucleus of the atom, which had grown out of the discoveries made some fifty years earlier that the atom was not immutable but mutable. Atomic physics had shown that the atoms too, had histories, and sent out trajectories of change over time in the form of packets of energy called radiation. Change, it seemed, was built into the very fabric of matter.

Some forms of change were, however, more natural than others.

Radioactive isotopes had been entering the earth's environment in ever-greater numbers, since the Trinity test and the detonation of the bombs over Nagasaki and Hiroshima in 1945. These were not the stable, "natural" isotopes that had been present in the earth from its origins. These were the by-products of atomic explosions, unstable radioactive isotopes with unfamiliar names such as strontium-90, plutonium, iodine-129, caesium-125, and tritium. They threw off dangerous packets of excess energy that destroyed the fragile cells of living creatures. Acute radiation sickness could follow immediately, or the effects could be delayed, manifesting years later as raised rates of cancer and genetic abnormalities in the offspring of those exposed.

Atomic testing increased throughout the 1950s as tensions between the United States and the Soviet Union rose. By the end of the decade, much damage had been done by radioactivity to living creatures in the environment. Exactly how much, no one knew. To answer this question, the International Atomic Energy Agency (IAEA) partnered with the World Meteorological Organization (WMO) to try to find out exactly how much man-made radioactivity was circulating, and where. In order to do so, they established just what Dansgaard needed—a global network for rainfall collection. Such global projects had roots extending back to 1853, when Matthew Fontaine Maury, the head of the U.S. Naval Observatory, called for an international system for monitoring the weather over the land and sea. In 1905, Léon Teisserenc de Bort proposed that the International Meteorological Organization (the precursor to the WMO) create a *Réseau Mondial,* or global network, for meteorological observations. Both had faltered under the weight of their own ambitions. Small differences in national observing practices added up to major inconsistencies in the data. Some countries, for example, took meteorological measurements at three-hourly intervals, while others insisted on two-hourly intervals. Differences in instrumentation, and varying guidelines for locating and reading instruments, compounded the problems. Even if data could be meaningfully compared, the sheer scale of the project was overwhelming and much data was left raw and unreduced.[10] More successful were a series of dedicated observing "years" which focused the efforts of scientists for a limited period of time. Two International Polar Years (held in 1882–1883 and 1932–1933) had laid the foundation for a more ambitious International Geophysical Year (IGY) in 1957–1958, which galvanized more than sixty-five nations, including both the United States and the Soviet Union, in a rare moment of scientific détente, into an eighteen-month frenzy of studying the earth from every imaginable angle.

In comparison with the IGY, the IAEA-WMO collaboration was narrowly focused, aiming solely to track the spread of tritium, a single radioactive element released by atomic testing. Soon, more than a hundred stations around the world were equipped to collect precipitation samples and send them to a central office in Vienna. The same water samples that they collected also contained—as did any water sample on earth—a distinctive ratio of oxygen isotopes. Dansgaard got wind of the project and seized another golden opportunity to piggy-back on the global reach and deep pockets of an international organization. All he needed was a few milliliters of the water to do his analysis. He mobilized a Danish contact on the IAEA and thereby hitched a ride on this great global rain hunt. It was not the first or only time that military matters would influence the direction of new science. As ever, Dansgaard proved adept at taking what he needed from such well-funded enterprises.[11]

The network spread itself as evenly as geography and politics would allow. South and North America were covered from Barrow, Alaska, in the north to the Falkland Islands in the south. Africa was speckled with stations, as were Europe and Australia. The great conjoined expanse of the Soviet Union and China remained an unrepentant blank. That left a large hole in the global precipitation network, but there were enough samples from elsewhere that a good understanding could be gained. Dansgaard was soon swimming in samples—a hundred per month (one from every station) at one point. The mass spectrometer machines—a fancy French one for measuring deuterium and the old standby from 1951 for the oxygen isotope—were run day and night.

To be useful, a thermometer must be consistent and give the

same reading for the same temperature no matter where it is. Water evaporates and condenses in complex ways as it travels around the globe. It wasn't clear that the processes that made cold air isotopically heavier in Copenhagen would hold true at every location. Sampled water had a "pre-history" of multiple condensations and evaporations which determined how much of a given oxygen isotope it contained. Would a consistent relationship between the percentage of heavy water and local temperature hold for water samples collected around the globe, each with vastly different pre-histories? The answer, with some caveats, was yes.

The IAEA-WMO data showed that it was possible to use the oxygen isotope technique to trace the "circulation patterns and mechanisms of the global and local movements of water," as Dansgaard put it in the paper he published on the results.[12] More than a decade earlier, Dansgaard had begun collecting rainwater in his back garden in Copenhagen. He'd shown that he could peer into the heart of a single storm using his mass spectrometer. Now he had shown that oxygen isotopes bound in water could reveal global patterns of evaporation and condensation, the flow of water molecules that moved heat around the earth and drove both ocean and atmospheric circulation. As temperatures fell at the stations whose ratio of oxygen isotopes he could measure, he found that there were more and more of the isotopically heavier atoms. The old correlation—between heavier isotopic weights and colder temperature—held in a variety of locations across the globe. He included a graph showing how the samples from a range of Pacific islands lined up, with the neat correlation between temperature and ^{18}O ratios holding across thousands of miles.

Dansgaard had now shown that the ratios of oxygen isotopes could be used as a thermometer, but he still wanted to know if they could be made into a time machine. The very radioactive fallout that had helped his progress, in that it had spurred the IAEA-WMO cooperation, now intervened. As a follow-up to his second Bubble Expedition, he was working on the problem of using a radioactive isotope of silicon, known as Si-32, to date ice, but the samples he'd taken from icebergs were all contaminated with fallout from Soviet tests. Dansgaard now turned to a peculiar and fascinating location—and project—that was literally underway on Greenland. Located on a portion of the island that lay directly on the flight path between the eastern United States with western Russia, Camp Century was an experimental U.S. Army station dug under the snow. Located inland just over one hundred miles (thus the name) from Thule Air Base, Camp Century became a unique piece in the puzzle of Cold War geopolitical strategy.

For the U.S. military, understanding the Arctic environment was critical to operating successfully within it.[13] One army official noted that although it was unusual for military researchers to have to "reach so far down toward the foundations of science as it has to do in this exceptional case," knowledge of snow and ice was so minimal that much basic research was necessary.[14] Camp Century was run by the Snow, Ice, and Permafrost Research Establishment (SIPRE), a Defense Department laboratory set up in 1948 to prepare the U.S. military should it need to mount an offensive across Greenland. Could heavy airplanes land on snow-compacted runways, or even floating sea ice, in order to supply stations along a planned network of fifty Arctic radar stations? Would it be possible to launch nuclear weapons from under the ice? Could a ground campaign be launched across the frozen expanse? Was it feasible to build and run a railway under the snow, to transport goods and men?[15]

By 1964, the U.S. Army had been actively trying to answer these fantastical questions for five years. They had built a futuristic sub-snow encampment on a scale never attempted before or since. Two hundred men lived in a series of tunnels below the snow, complete with a 4,000-book library, laundry, mess hall, barber shop, and hospital, and liberally supplied with hearty cuisine including steaks, green beans, and mashed potatoes. It was all powered by a nuclear reactor located only 300 feet from the men's living quarters.

Dansgaard was largely unimpressed by Camp Century's operational ambitions (he noted the folly of attempting to build a railway under the snow—it had started to buckle and bend under the pressure of the ice as soon as it was begun), but it could provide him his much-needed untainted snow, preferably snow which was just

FIG. 7.4. The thermo-drill set up at Camp Century, 1964. Dansgaard never saw the rig during his visit.

older than the tests which had released Si-32 into the atmosphere. The depth to which Camp Century was dug had exposed just this sort of snow, so he set out in the summer of 1964 to collect some samples. He duly gathered the ice he needed, noting the weird life under the ice, with its mix of American cheeriness (drinks cost just twenty-five cents regardless of size) and brutal extremes of cold and pressure.

The few days he'd spent at Camp Century he'd been busy with his own work. Just meters away, on the other side of a wall of snow and ice, an enormous drill was coring deep down into the thick cap of ice which covers Greenland. Dansgaard left Camp Century without ever seeing the rig, which was a military secret. Over the course of half a dozen years, the Americans drilled deeper into the ice than anyone had ever managed to go. The practical challenges of drill-ing so deep were substantial, with the walls of any drill-hole sub-ject to enormous pressures. The tool that did the job was a special thermo-drill that simultaneously melted its way through the ice and preserved a core of frozen ice uncontaminated by meltwater. It was so expensive that only the U.S. military could afford it. By 1966, the drillers had hit bedrock, 1,390 meters beneath the surface of the ice.

Once Dansgaard got wind of the long core, he knew it would be just as valuable for studying Arctic climate as the IAEA-WMO samples had been for studying the water cycle. Dansgaard had proven his ability to use isotopes to study samples of both ice and liquid water to study climate in the present and the past, and it wasn't long before he'd persuaded the right people to give him some samples from it. Though neither he nor the U.S. military knew it yet, by far the most enduring and valuable product of six years and countless dollars would be the result of what Dansgaard and his colleagues did with these samples. What he did would generate data that showed just how varied the earth's climate had been in the past, and, just as importantly, how abruptly the climate could change.

As early as the 1820s, Joseph Fourier had recognized the important role played by the atmosphere in trapping the sun's heat, but he had never imagined that humans might one day materially affect the amount of certain key gases in the atmosphere. John Tyndall had rendered quantitative just how certain molecules absorbed heat, but he too had not considered the possibility that humans might sig-nificantly affect the earth's atmosphere in the future. It was only in 1895 that a Swedish physicist, Svante Arrhenius, had made the first calculation about the effects of an unnatural (or forced) increase in carbon dioxide in the earth's atmosphere, due to the action of humankind. His prediction, which looks incredibly prescient today, was rejected by other scientists as implausible. In 1938, an English engineer named G. S. Callendar had performed additional calcu-lations on what a doubling of carbon dioxide would mean for the average global temperature.[16] Again, his contemporaries considered his findings unlikely. It was only in the 1950s that work by Roger Revelle and Hans Suess brought sustained attention to the effects of carbon in the earth's oceans and atmosphere. Responding to their work, the geochemist Charles Keeling established a measurement station atop Hawaii's Mauna Loa volcano to record the percentage of carbon dioxide circulating high in the atmosphere, where it was most evenly distributed. Almost immediately, his device began to reveal a gradual and inexorable rise in the amount of CO_2 in the atmosphere as a result of the burning of fossil fuels that released carbon deposited millions of years ago.[17]

This was the moment, as it were, when the past met the future. While a handful of researchers had foreseen the effects of industrial activity on the climate of the entire planet, their work had not found a receptive audience. It was only in the late 1950s that a particular combination of researchers, instruments, questions, and anxieties came together to prompt sustained and productive research into the nature of the effects of human activity first on the atmosphere and, more recently, on the entire global system. It was also when the term *climate* shifted from meaning something primarily geographical to a temporal concept connoting change. While men such as Köppen and Hann had worked to "modernize" climatology and render it globally comprehensive, they had remained committed to an understanding of climate as essentially stable on human timescales (i.e., for decades or longer). It was only with the indications of the rapid rise in carbon dioxide, and a growing realization of what that could mean for the climate of the entire planet, that a new kind of climate science came into being, one which emphasized temporal changes on a global scale.[18]

Gradually, these findings on rising carbon dioxide made their way into public sphere as the findings of individual researchers emerged and certain scientists, notably Roger Revelle, began to speak loudly enough to require responses from those in power. Further investigations were then launched into what the practical impacts might be. In 1965, President Lyndon Johnson convened a panel on environmental pollution. The panelists were tasked with evaluating the direct environmental effects of industrial pollution on air and water

quality but decided to include a special report on an "invisible pol-
lutant," atmospheric carbon dioxide. Chaired by Roger Revelle,
the five-man subcommittee on carbon dioxide included Charles
Keeling, the man responsible for the CO_2 measurements on Mauna
Loa, as well as geochemist Harmon Craig, meteorologist Joe Smagorinsky, and a young geologist and geochemist named Wallace
Broecker. Referring to earlier work by Svante Arrhenius and T. C.
Chamberlin on the effects of rising carbon dioxide on climate, the
subcommittee reported on the five years of records then available
from Mauna Loa (and another at the South Pole) that revealed a
1.36 percent rise in the carbon dioxide content of the atmosphere.
Based on estimates of fossil fuel consumption in the past and projected future use, they guessed that by the year 2000 there would be
roughly twenty percent more atmospheric carbon dioxide than pre-
industrial levels. Considering the extra heat that would be trapped
by this extra carbon dioxide, they estimated a rise in the average
temperature of the earth's surface of between 0.6°C and 4°C.

The authors of the report readily admitted that these estimates
were based on many simplifying assumptions. Despite the impos-
sibility of making precise predictions, they were still confident that
by the year 2000—then thirty-five years in the future—the rise in
carbon dioxide would be significant enough to produce "measur-
able and perhaps marked changes in climate," including changes in
temperature.[19] Given the scale of the changes, the effects on human
beings could be "deleterious," and the authors suggested that it was
important to explore the possibility of "deliberately bringing about
countervailing climatic changes." The future was close in which it
might be advisable to spread reflective particles over large portions
of the upper ocean, or to alter cirrus clouds high in the stratosphere.

As word began to spread throughout the scientific community
that carbon dioxide was rapidly rising, it became increasingly clear
that it would be necessary to understand how Earth's climate operated at the largest possible scale. Given the complexity of the task at
hand, the scientists working in this area started as simply as possible.
They created climate "models," sets of equations that described a
few basic features of the climate system.[20] They used these toylike

models to play with the climate system, testing what changed as they varied a small number of inputs, such as the amount of carbon dioxide in the atmosphere or the amount of radiation being introduced into the system. Since these models were designed to understand the impact of a global metric—the amount of carbon dioxide—they also generated a global output: the global surface temperature.[21] This single number usefully reduced the otherwise overwhelming complexity of the planet to something even a child could grasp. It also suggested that something called the "global climate" existed. This so-called global climate was in many ways a useful fiction, born of often necessarily crude averages. The global average temperature did not—and cannot—exist in any one place. It was an imaginary tool for grasping the earth in one gesture, a useful bit of reduction that simultaneously enabled tremendous insights into the working of the global climate system and elided its complexity.

To check if these models were realistic, these early modelers needed to compare them with real data from the earth. This demand for global temperature data prompted scientists who worked with such data—among them climatologists working at the Climatic Research Unit at the University of East Anglia (UEA)—to start producing the first estimates of the average global temperature, based on measurements collected since roughly 1850, when reliable instruments were first introduced.[22] Unlike global climate models, which are easy to build as long as equations are kept simple and few data points are entered, there is no easy way to generate a global average temperature. The only way to do so is to take as many measurements around the planet as possible and find a way to combine them that takes into account all the local variations, and gaps in coverage, that would otherwise queer the result. Indices of temperature readings had been compiled, beginning in 1938, by G. S. Callendar, by Mikhail Budyko, and by J. Murray Mitchell Jr. These averages relied on readings from the Northern Hemisphere. It would be decades before data from the oceans was incorporated into these putatively global averages, and even longer before observations from remote polar regions were included.[23]

The notion of an average global temperature changed what it

meant to study climate. It did not, however, spell the end of climatology as it had been long practiced by inheritors of the Humboldtian sensitivity to location, men such as Köppen, Hann, and Hildebrandsson. Indeed, Hubert Lamb, a climatologist with a sophisticated historical and geographical awareness, had founded the Climatic Research Unit and remained an important figure in the field.[24] But global temperature indices did help bring about an important shift in what it meant to study climate that ran counter to the geographically oriented strain of climatology Lamb had advocated. Once the planet was considered as a machine for producing average temperature, local or even regional variations became, if not irrelevant, then subsidiary to the aims of global climate modelers. Here, then, was the birth of a new science of climate—different than climatology—that would concern itself not with the places of the planet but with its past and its future, with the mechanisms that prompted changes that could be measured on a global scale.

Far from abandoning the past, for the scientists studying the effects of carbon dioxide, the newly uncertain future they were facing made the earth's history an even more precious resource. If carbon released by human activities had the potential to dramatically alter climate, what had happened in the earth's past potentially became the key to the future. Only by understanding the natural variability in the earth's past—when it had been warmer, or colder, and why—would it be conceivable to predict what lay in the future. Prediction had been the long-desired mantle of scientific maturity that the meteorologists of the nineteenth century had sought. In the decades following World War II, it seemed as if the new field of climate science might finally be able to deliver on the promise of predicting not only the weather but the earth's climate as well. At the same time, a small group of scientists grappled with the realization that the climate was already responding rapidly to changes caused by human beings. In this way, the ambition to predict the future climate was always entangled with a newfound awareness of just how variable the climate was, and just how much power humans had to perturb this already unstable system.[25]

❋

FIG. 7.5. A storage freezer full of samples from the Camp Century ice core at the Cold Regions Research and Engineering Laboratory in Hanover, New Hampshire, 1965.

While these changes were occurring in the discipline of climate science, Willi Dansgaard kept working to gain access to the ice drilled by the U.S. military at Camp Century in Greenland. By 1967, he had prevailed. In that year, ice that had been extracted from the interior of Greenland with great difficulty and expense, and amid much secrecy, was sent in two-meter sections to a New Hampshire government laboratory devoted to studying cold, the U.S. Army Cold Regions Research and Engineering Laboratory (CRREL). Soon after, a Danish colleague arrived in New Hampshire to collect eighty-six samples to be taken back to Denmark for analysis in Dansgaard's mass spectrometer. He finally had a chance to try out the isotopic time machine he'd been dreaming of.

He now had samples from the entire length of the core. Dansgaard and his team analyzed almost 1,600 different pieces from the nearly mile-long core in their mass spectrometer. When it had been reduced and plotted, the data showed up as a long and squiggly line extending back in time. The core had produced better results at a much finer resolution than anyone had expected.[26] The sequence of rises and falls in temperature indicated by the changing amounts of heavy isotopes in the sample were complex, but there seemed to be underlying cycles. The detail revealed in the ice core, especially for the most "recent" 8,300 years, was stunning. It was possible to trace annual temperatures, including seasonal summer/winter variations, as if reading the rings of a tree. The most stunning number was 100,000, the number of years the ice record stretched back. It was by far the longest record of the earth's climate that had ever been obtained at such a high degree of resolution (the sediment cores taken from lakes, though delivering longer timescales, were much less finely resolved, with a thousand years squeezed into a single centimeter). Ice cores, in contrast, laid out annual bands at a rate of roughly 50 years per meter. Easy to see, and count, they made the past legible in a way it never had been. In a splashy 1969 paper on the "one thousand centuries of climatic record" that appeared in *Science*, Dansgaard and his team identified a series of cycles, or climatic oscillations, within the results. Roughly every 120, 940, and 13,000 years, they suggested, there were regular changes in the climate.[27]

The first thing Dansgaard and the team did was compare their findings with other clues about the earth's past temperature. Around the same time that Dansgaard had been investigating the nature of the isotopic ratio of oxygen in rainwater, other physicists had been using the same ratio to turn muddy cores taken from the ocean floor into thermometers in their own right.[28] If they matched, that was a good sign that the ice core was telling them something meaningful about changing temperatures in the past. Otherwise, there was a chance the ice-core results were erroneous or represented temperature changes limited to Greenland. Dansgaard and his collaborators included a figure in their *Science* paper comparing their results with past temperatures derived from other studies—of ancient pollen from Holland that extended back 80,000 years, Pleistocene deposits that stretched back nearly as far, and deep-sea sediment cores. Two facts immediately stood out when the four traces of climatic varia-

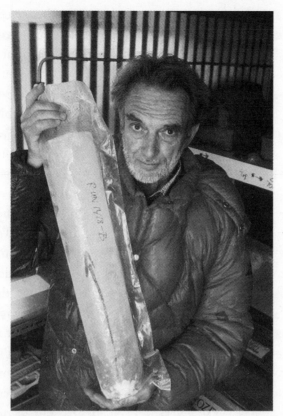

FIG. 7.6. Willi Dansgaard in the freezer with an ice core, 1994. Credit: Centre for Ice and Climate, University of Copenhagen.

tions were placed side-by-side. First, the broad outlines of the four curves, each obtained by independent means, did indeed match up. Second, the detail offered by the ice-core study was phenomenal. The other traces looked like a child's messy scrawl compared to the minutely saw-toothed edge of the ice-core data.

While Dansgaard and his coauthors stressed that their curve was "primarily valid for the North Greenland area," they ended their paper by noting the uncanny correspondence between the readings. The correspondence gave credence to their results, but it did something even more important: It gave strong evidence that major changes in past climates had occurred on global, not regional, scales, and that they had occurred much more rapidly than anyone had ever imagined. Their dry language couldn't conceal the excitement they felt about a technique that provided "far greater, and more direct, climatological detail than any hitherto known method."[29] What this detail revealed was the potential for abrupt changes in the climate of the planet.

Dansgaard's technique, for all its ingenuity, would have been much less valuable without the remarkable length of ice on which to work. The drill at Camp Century had made it possible along with the unique skills and sheer stubborn-mindedness of the team that operated it. Their success made the drill team hungry for more ice. Soon after hitting bedrock at Camp Century, they embarked on another great drilling expedition, to Byrd Station, a U.S. research station established in Antarctica during the 1957–1958 IGY. Within two years, they'd managed to pull another core from the ice, which went back almost as far in time, and with the same degree of detail, as the Camp Century core. Once again the results were compared, and once again they matched up.

The changes that were preserved in the ice at the top and the bottom of the globe seemed to be the traces of dramatic climatic shifts (up to 8°C) that had affected the entire planet in the space of just a few decades. No one had ever imagined climatic change of such magnitude could happen so quickly.[30] Even after the publication of the paper, the meaning of the curves that Dansgaard had produced was not readily apparent. Though Dansgaard and his coauthors knew they had created a powerful new way of looking into the earth's past, the community of researchers interested in understanding the earth's past—and the implications it held for the future—did not yet fully understand what the ice-core traces meant. Decoding the jagged curve fully would take an individual with a mind that could range, as had James Croll's, across disparate data and a tremendous variety of both spatial and temporal scales, to understand fully the implications of the rapid changes revealed by the ice core.

Wally Broecker was just such a person. Trained as a geochemist with a special interest in ocean isotopes, he had served on the 1965 subcommittee on atmospheric carbon dioxide along with Roger

Revelle and Charles Keeling. In 1966, he had noticed "an abrupt transition between two stable modes of operation of the ocean-atmosphere system" in a set of cores taken from the floor of the deep ocean, stretching back 200,000 years.[31] Primed to see fast transitions, when Broecker read Dansgaard's 1969 Science paper, he immediately noticed how abrupt the changes revealed by the cores were. He realized that the cycles that Dansgaard and his colleagues had identified in the past could also be used to predict future temperatures with much better resolution than the crude relationship between atmospheric CO_2 and temperature that he and his fellow subcommittee members had had to rely on in their 1965 report. The timing of the most recent warming bump in the small-scale cycles reported by Dansgaard suggested that another such warming period was due very shortly. Drawing on the historical record and thinking along the same timescale of decades that the Johnson committee report had established, Broecker extrapolated to the very near future. His paper, published in Science in 1975, was titled "Climate Change: Are We on the Brink of a Pronounced Global Warming?" Today, it is a very familiar headline. At the time, it was not. While Broecker was not the first to use the phrase, his Science article put the term into wide circulation. His paper has since been celebrated as a landmark in the history of global warming. Its thirty-fifth anniversary was marked, and Broecker (to his considerable consternation) was dubbed the "father of global warming." At the time, however, the adjective global was just as significant as the idea that the climate might be warming. Global change—in any direction—was the signal that Dansgaard's cores had revealed. Precisely what kind of change remained, as Broecker's title had suggested, was an open question.

While Broecker's prediction came true, the inference on which it was based turned out to be groundless. He had taken what he himself called a "gigantic intellectual leap" when he'd assumed that the small-scale cycles in the Camp Century core would also typify the globe. He readily admitted as much. "My prediction was based on the false premise that Dansgaard's record typified the globe. In reality, it typified only the northern tip of Greenland." Subsequent ice cores from different locations didn't reveal additional 120-year cycles. Broecker's prediction of renewed warming came true, but not because of any underlying global 120-year cycle revealed by the Camp Century core.[32] It was a lesson in the challenges of predicting changes in climate that scientists would learn again and again. The global system was a complicated beast, and no single heartbeat could explain the changes it had undergone in the past—nor those it might experience in the future.

Thanks to Dansgaard's idea and the developments in isotopic analysis on which it relied, ice cores became, in Dansgaard's poetic phrase, frozen annals of the earth's past. Ice cores were perhaps the most sublime of paleoclimate records, with their evocation of the cold poetry of ancient snows, but there were many other, equally amazing, if less haunting, records of the past, some of which stretched even further back in time. In the hunt for more clues, paleoclimate scientists analyzed layers of mud and sediment taken from the ocean floor that preserved the shells of ancient sea creatures called foraminifera that had built their shells out of isotopically diverse elements. They bored into ancient trees, counting the bands of tree rings that documented the changes experienced by trees that had lived since well before Christ. They studied pollen deposited on winds that had blown thousands of miles, leaving a record of which plants had flourished and therefore an indication of the climate conditions. Getting a tree to give up its secrets took different skills and required different assumptions than finding a way to make ancient mud speak of long-past climates. The strength of the signal varied, as did its precision. The old mud was, by its very nature, blurrier than the rings of ancient trees, but while the trees could be used, en masse, to go back 11,000 years, the mud stretched more than 1.5 million years.

It was a simultaneously stimulating and contentious time. There was a heightened self-consciousness of the need for interdisciplinarity, but the challenge of working across disciplinary boundaries remained. An important conference took place at Brown in 1972.

Dansgaard, who did not attend, sent in his results, which were included in the proceedings. Despite the productivity of the meeting, the tension was also evident. Two kinds of researchers found themselves in a room together for what felt like the first time. On the one hand were researchers who searched for "climate analogues," past episodes that looked like the present climate and which could provide some indication of what was to come. On the other were scientists who wanted to determine the physical causes of those patterns—to get underneath the patterns and explain their origins. It was unclear which direction the field would go. The editors of the volume that came out of the conference thought it wise to allow both strands to flourish. "These two different approaches to research on climate change must run their own courses until such time as the validity of general theories is well demonstrated."[33] Time would tell which approach would win.[34]

Much was up for grabs. Where Louis Agassiz had posited a single ice age, and James Geikie and James Croll had argued for multiple ice ages, to researchers like Dansgaard and Broecker, the paleo-climate data indicated that climate was, in a certain sense, always changing. Seen in the context of a temperature curve that, however tentatively, extended back some 100,000 years, any climatic stability looked temporary. The old idea of a stepped climate, extending back to Lyell, was abandoned. In its place was a new way of thinking about climate as something that was, in a certain sense, undergoing constant change. It was becoming increasingly clear that, as Barry Saltzman put it, "there has always been so much variability of the climate that it seems doubtful that we can ever speak of a single climatic norm for the Earth. Judging from the past, the climate we are experiencing today is almost certain to be transient, giving way to something different in the future."[35] The new way of seeing the earth's past—and by extension its future—as a constantly changing system represented a massive reorganization in the way scientists thought about the earth.

When Broecker had warned in 1975 of the possibility of global warming, he did so against a backdrop of fears about not global warming but global cooling. A string of colder than normal years

prompted some scientists to worry that another ice age was imminent, as the natural global climatic cycle entered a cooler stage. The
potentially catastrophic effects on the global food supply prompted
front-page-level anxiety in 1972 against a backdrop of Soviet grain
crises. Climate was changing, and which way it might change seemed,
for a moment at least, to be up in the air.

The abrupt warming episodes revealed by Camp Century and
other ice cores drilled in the late 1960s and 1970s came from the oldest ice, which sat at the bottom of the ice sheet. As a consequence, the
records were often smeared or distorted by the pressure of the ice as
it dragged along the bedrock—just where it was most critical to see
it clearly. What was the nature of the earth's changing climate in the
past? How variable had it been? What patterns in its changes might
be found? More ice cores were needed in order to solve a mystery
only just coming into focus. Cooperation between American and
European consortium projects proved difficult to secure. The science around ice cores was so exciting that each nation wanted a
bite of the apple. In 1987, two different proposals emerged to drill in
central Greenland. Far from discouraging this duplication of effort,
Dansgaard recognized it as an important way to validate the results
of each individual core. Much as Piazzi Smyth's stereo-photographs
had done more than a century earlier, the ensuing ice-core projects,
named GRIP and GISP2, would provide a check on each other.[36]
Both projects got underway in 1989, just twenty miles apart from
each other, in the very center of Greenland, where the ice would
be thickest and hopefully the least distorted by sliding. While they
failed to produce the really old climatic information researchers had
been hoping for, these cores produced stunning results confirming
the abrupt changes that earlier cores had suggested.

The data from these new Greenland ice cores further galvanized
Broecker's thinking. Once he'd been primed to see the changes
in the Greenland cores, he began to look for similar changes in
many other paleoclimate proxies. "One after another, other archives
showed the same thing," recalls Broecker. "Greenland led the way.
No one would ever have thought that that had happened. If Dansgaard hadn't, we'd probably have had a much harder time realizing

FIG 7.7. The three musketeers of ice core research: Willi Dansgaard, Chet Langway of the CRREL, and Hans Oeschger, circa 1980.

these things."[37] Broecker named the abrupt changes captured in the Greenland cores Dansgaard-Oeschger (D-O) events because they were found both in ice cores that the Dane had helped unlock and for which Hans Oeschger had developed techniques for analyzing the gases trapped within. With additional confirmation from ice cores drilled in Antarctica, Broecker's idea that many of the cycles in Greenland's cores represented global changes, and that the D-O events were startlingly abrupt cycles of change, seemed to be confirmed (even if the short 120-year cycles were never found anywhere else).

The cause of these sudden events, both the D-O events and a series of even more dramatic shifts called Heinrich-Bond cycles, is still debated. Broecker's best guess, and still a prominent hypothesis today, was that a shift in the way the ocean circulates caused by an influx of melted ice in the North Atlantic could have caused a dramatic drop in temperature. While the currents in the top 100 meters of the ocean are driven by winds, down in the depths of the ocean it is the density of water that determines its motion. When water at the poles cools, gets saltier, and sinks to the bottom of the ocean, it sets up a chain reaction that draws in warmer, fresher water to replace it. The cycle transfers huge amounts of energy around the globe and, Broecker realized, could be responsible for a very fast and dramatic shift in the climate if something were to disrupt it.[38]

The lessons of the ice cores were coming through loud and clear. Change came first, and from it all else followed. This new understanding of the planet, based on ice cores and other paleo sources, began to make its way—like the melting of ice in a glacier—into the deepest recesses of the growing network of national and international bodies dedicated to thinking about the earth as a system. Systems thinking and change-thinking went hand-in-hand—the changes revealed by the ice cores needed systems of interlocking mechanisms like the ones Broecker explored to explain them.

At the same time, a growing environmental consciousness made the changes happening in the present thanks to the activities of humans feel increasingly urgent. Not only did human activity seem to be increasing in its pace, but it was increasingly evident that it had the potential to effect change on a planetary scale. Global change became the organizing principle of a series of influential workshops held in the early 1980s. This was the moment when all roads seemed to lead, as NASA put it, on a Mission to Planet Earth. Inspired in part by the new availability of observations of Earth taken from space in the 1960s and 1970s, these meetings captured the feeling of the time and would have a lasting influence on what came next.

"The earth is a planet characterized by change," declared the participants of a NASA workshop on "Global Change: Impacts on Habitability," held at Woods Hole in June 1982, "and has entered a unique epoch when one species, the human race, has achieved the ability to alter its environment on a global scale."[39] The long run of resource extraction and exploitation that had fueled most of human development was, it seemed, coming to an inevitable end. "The next valley," warned the participants, "is now occupied." The need for working across disciplines was paramount. "We have now reached a point where the boundaries of each discipline are overlapping, and the next step forward can only come from an interdisciplinary research program."[40] The next summer, another workshop took place, and with it came another impassioned plea for whole-earth thinking: "The earth is changing even as we seek to understand it," reported the participants of a meeting funded by the International Council for Science to investigate the establishment of an International Geosphere/Biosphere Program, "in ways that involve the interplay of land and sea, of oceans, air, and biosphere." Only by seeing the earth as a "single system" could there be any hope of understanding the problem.[41] That statement of the need for a new way of seeing the earth had behind it the intellectual heft of Roger Revelle—the man who had first raised the alarm about increasing concentration of carbon dioxide.

The most influential report, however, came from NASA's three-year-old Earth System Sciences Committee, chaired by Francis Bretherton, the theoretical oceanographer who had helped guide the design of the Mid-Ocean Dynamics Experiment. Recent developments, Bretherton wrote, had "converged to reveal to us—indeed, to force upon us—a new view of the Earth as an integrated system, whose study must transcend disciplinary boundaries."[42] At a meeting of the project's modeling group at the incongruous location of the Snow Bunny Lodge, in Jackson Hole, Wyoming, team members alternated between skiing and sketching out a diagram showing "how all the pieces of the planet work."[43] One of the participants, caught up in the intensity of the discussion, momentarily forgot that the diagram in progress was being projected on the hotel room wall. He corrected an equation with marker directly on the wall, at some cost to NASA, who footed the bill to repaint it.

The Bretherton diagram, which took its name from the chair of the committee that produced it (though Francis Bretherton did not himself play a direct role in creating the diagram), was an attempt to capture the complex interactions between the various elements of an earth system. These feedbacks showed how each component of the system could affect and be affected by each other. Together, these feedbacks created the variability in climate that Dansgaard's ancient ice cores had revealed. Notably, the diagram brought both the biogeochemical and the physical aspects of the earth system together—the living ocean and land and the swirling air and water upon which that life depended. Ocean dynamics was connected to both atmospheric physics and marine biogeochemistry, for ex-

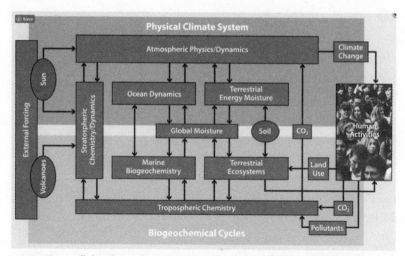

FIG. 7.8. The so-called Bretherton diagram, an influential product of NASA's 1986 Earth System Sciences committee, showing physical and biogeochemical systems linked together. Human activities are represented by a single box to the right.

ample, while terrestrial ecosystems drew from the soil and global moisture while affecting tropospheric chemistry. At the center of it all sat global moisture. It was water that made Planet Earth, uniquely in the solar system, it seemed, a planet of change.[44]

While it wasn't the first time anyone had considered it important to link the living earth with the physical earth, the diagram appeared at a moment of elevated self-consciousness. The 1960s and 1970s were crucial decades in the development of an awareness about climatic change that had affected the entire planet over a mind-boggling array of timescales, from decades to hundreds of thousands of years. Paleoclimatology was a broad church, bringing ice core, trees, mud, and pollen studies under the same roof. Broader still was the new science of climatic change which brought geologically minded practitioners seeking to understand the past ice ages together with more physically and chemically oriented earth scientists who were working to predict the effects of a human-induced rise in CO_2.[45] It may not be too much of an exaggeration to say that the big discovery of twentieth-century climate science was not that humans had the ability to change the climate but, as the tools of paleoclimatology were showing, that the climate itself was always changing.

In this context, it seemed more important than ever to try to show, as the Bretherton diagram did, how living things affected and were affected by the movement of water and heat through the system. To that end, the diagram also included a single box representing human activities, which affected the climate system via land use, pollutants, and carbon dioxide, and which was, in turn, affected by the outputs of the system in the form of climate change and terrestrial ecosystems. This little box represents a significant milestone in global change research. To be sure, it is as reductive a representation of human activities as it is possible to imagine. By squeezing all of human affairs into a literal black box, humanity seems to function as little more than a cog in the great mechanism of the planetary climate system. While this may have indicated a certain humility about human impacts on the planetary scale, it also demonstrated a strikingly blithe attitude toward the complexity of human activity. Looking back, it is hard not to read this box as an indication of the naiveté of those who came up with the diagram, vis-à-vis the social sciences. And yet, crude as it was, the inclusion of this box represented a sea change in the way climate scientists understood their science. By including human activities within the earth system, these scientists were not only acknowledging that human beings had the potential to alter the planet in extremely significant and consequential ways. They were also indicating that the tools of climate science alone would not be enough to describe the nature of the earth system, much less to design programs that could limit or reverse a warming trend.

The Bretherton diagram became famous among the influential group of scientists and administrators working within or alongside earth system programs at NASA, and as part of the International Biosphere/Geosphere Program established by the International Council for Science in 1987.[46] Most fundamentally, the diagram conveyed in simple, graphical form a vision of the planet as a system—an engineer's Earth made up of interlocking parts. The implication of the diagram was a pragmatic one: If the earth was a system, each part of that system could be treated as a modular element. In this way, the Bretherton diagram reduced the otherwise overwhelming complexity of a global climate system into a diagram that could be used, as an electrician might use a wiring chart, to diagnose crucial locations in the system—or tipping points—upon which study and intervention could be focused.

This vision of the planet was similar to Stommel's ocean—a complex machine with many moving parts, but nonetheless a machine in which parts could be profitably studied independently of one another. It was a pragmatic and mechanistic vision of the planet suited to engineers. What the Bretherton diagram did not do was promulgate a vision of a living planet, or a planet which, Gaia-like, constituted a living being and in some sense acted to conserve life. Instead, it captured the vision of the planet as seen by a group of engineers and scientists who had been empowered by the national and international agencies they worked for to try to solve a problem. That problem was to understand the natural variability of the earth in time to be able to address the additional problem—crudely but significantly captured in the single box—of the effect of human activity on the system. For all its reductionistic framing of human affairs, the report and associated diagram were a claxon announcing the existence of a new field in which human activity was inextricably linked with complex climate processes. The earth was not one thing, nor was it many things. It was, instead, "an integrated system of interacting components," of which human beings were nominally, if only crudely, seen as forming a part.[47] This was the beginning of a new field that called itself Earth system science. Each of the key words in this phrase was loaded with implications for how NASA, which had already positioned itself as a key knowledge-producing organization of this new science, saw the future of the sciences of the earth.

The vision of the earth generated by the *Apollo* space missions was not merely, or even primarily, a vision of unity and shared responsibility for a fragile living earth which Stewart Brand promulgated with his *Whole Earth Catalog*. The image of Earth from space was, in this sense, a planet calling out not for salvation or the development of a shared planetary consciousness, but for management. What was being managed was a system characterized by change. The discovery of change as an integral feature of the earth's climate system was simultaneous with the development of methods for engineering or managing that change. To put this another way, the discovery of anthropogenic climate change was simultaneous with the discovery of natural climate change. The ability to recognize abnormal change was predicated on the vision of past natural change that the paleoclimatologists—not least the ice-core scientists—revealed. And the structure that scientists from a range of disciplines used to combine and compare these newly available observations was that of Earth system science. Change was built into this system, and the subtitle of the report was a "program for global change," punning on the need for action to address humanity's role in perturbing this system as well as the need to study change on a global scale. "The people of the Earth are no longer simple spectators to the drama of Earth evolution," went the story, "but active participants on a worldwide scale."[48]

The Bretherton diagram—and the fledgling discipline of Earth system science that produced it—today speaks poignantly of both the hubris of thinking it might be possible to "solve" the earth system and the humility that comes from confronting the scope of the problem. "The study of the Earth is on the verge of a profound transformation," proclaimed Bretherton's report, but so, too, was the planet itself: "human activity is now causing significant changes on a global scale within the span of a few human generations."[49] The implicit question raised by the report was whether human beings could catch up with themselves before it was too late.

8

CONCLUSION

Arriving at the Bretherton diagram after spending time in the company of the scientists I have profiled, it is possible to see it as an amalgam rather than a singular thing. In a quite literal sense, one can locate within it (and its accompanying text) much of the work with which this book has been concerned. Here, for example, is the Tropical Rainfall Measuring Mission, NASA's first mission to measure rainfall on Earth, which Joanne Simpson was recruited to run in 1986, the same year that the diagram itself was published. Here too is a mention of the World Ocean Circulation Experiment (WOCE), the global project for which Henry Stommel had laid the conceptual foundation and about which he was deeply ambivalent. Also present is Gilbert Walker's Southern Oscillation, whose erratic occurrences were still, in 1986, unpredictable (and remain so today), and which exemplifies the importance of studying climatic processes on a global scale. Less explicitly but perhaps more fundamentally, the climatic roles played by clouds and water vapor remain a subject of pressing mystery in the Bretherton diagram just as they did for Charles Piazzi Smyth. And here too is the need to understand the way ice moves and changes shape: whether, as Tyndall had also wondered, it slides on liquid water at its base, and how, in a warming world, its motions might lead to the detachment and melting of the massive West Antarctic ice sheet.

The lives and work of John Tyndall, Charles Piazzi Smyth, Gil-

bert Walker, Joanne Simpson, Henry Stommel, and Willi Dansgaard were lived, as we live our own lives, in a stream of constantly
shifting desire, intention, and chance. As Tyndall felt so acutely, a
misplaced step in the Alps could have brought him tumbling down.
Piazzi Smyth's life might have been very different if he hadn't rashly
resigned from the Royal Society. Gilbert Walker had a nervous
breakdown from which he recovered, but what if he hadn't? These
counterfactual stories are useful as reminders that the contingency
of individual lives has influenced the creation of what might otherwise seem to be a natural object—the system of the earth, the vision
of the globe.

Both Tyndall and Piazzi Smyth demonstrated that it was possible and sometimes necessary to make knowledge about the planet
alone. The nature of that knowledge, and what it meant to be alone,
was negotiable. When he was on top of Tenerife, Piazzi Smyth was
imaginatively supervised by a congregation of his scientific peers.
Tyndall often climbed "alone" in the Alps in the presence of guides.
Making reliable knowledge out of an amalgam of field and laboratory work was never straightforward. Gilbert Walker, in many ways
a useful anomaly here, sat at the center of a web of imperially enabled
statistics like a calculating spider. His failure to predict the monsoon
is less significant than his success at showing how certain powerful modes of calculation depend on unimaginably large networks of
observations, networks which are today even more important, and
arguably even less visible, than they were for Walker. Understanding how such hidden numbers produce highly visible and influential knowledge—such as the global temperature index—is critically
important.

A full history of our understanding of the planet cannot be told
solely from the perspective of individuals. Indeed, since World
War II the sciences of climate have become bigger and bigger, and
the role played by any one individual smaller and smaller. This was
the change that Stommel foresaw and lamented. As he saw it, it
entailed the loss of freedom that he considered necessary for solving the big conceptual problems. Joanne Simpson encountered the
same paradox when she sought to build alliances with government

funding bodies in order to do the science she really wanted to do. Cloud modification, much less hurricane modification, was never going to be a solitary endeavor. Similarly, no ice cores get drilled without serious money being expended. Willi Dansgaard had to find a way to harness the budgetary and logistical might of national governments. Piazzi Smyth and Walker, too, both relied on extensive networks to provide them with the necessary equipment and authority to do their work. Realizing that individuals cannot function alone does not mean that individual lives no longer matter to the telling of history. By looking at the interaction between individuals and institutions—between energy at different scales of the system—we come closest to understanding how the system works.

The tools with which scientists study—and then describe—the planet in global terms are influential. They generate knowledge that contains within it both assumptions about who gets to know—who has the training and skill, the moral authority, and therefore the trust to do so—and what that knowledge is good for. This book has conveyed the stories of how scientists have created tools for global knowledge and what was consequential about those tools. While my protagonists are almost all English speakers and nearly all men, they belong to different times, different places and, perhaps most challengingly, different disciplinary histories. I have chosen to do this deliberately, as I wished to show how the thing we today refer to quite casually as climate science is an amalgam of different ways of knowing the earth. This is, in one sense, a good thing, a source of resilience, in that it offers multiple pathways for generating knowledge. For decades, calls for the need for interdisciplinarity have been common, sometimes more strident than others. Despite this, truly interdisciplinary working remains elusive across the natural and social sciences. A recent meta-analysis of twenty scientific assessments of global climate science noted that "only a fifth of the case studies analysed attempt to integrate practical elements [or] consider socio-economic and geophysical aspects across spatial scales."[1] And yet, as this book makes clear, climate science was always interdisciplinary. For better or worse, there never really was a singular discipline of climate science.

Global visions are necessarily made up of unglobal things—individuals, places, moments in time. This is, in itself, neither a bad nor a good thing, but it is a fact about which it is important to be aware. The concept of global knowledge is a powerful one. It may be one that we feel we need today, but this does not make it either neutral or natural. All our global visions are, like the visions described in this book, the products of individual minds working in particular places at particular times—histories that might have turned out differently. Put differently, the same Earth is there for all of us, but, to use a phrase Tyndall might have appreciated, it wears many veils. The tools with which we pull back the veils go, in this book, by the names of the various scientific disciplines into which humans have divided the study of the planet: geology, physics, astrophysics, cosmic physics, atmospheric physics, meteorology, oceanography, paleoclimatology, climate science. Those disciplines organize the methods for thinking and doing science, and in this way they, too, determine what can be known by the people who work within them. There are circles within circles of structure and chance, of painstaking preparation and the unpredictable contingencies of the day. Together, it adds up first to individual lives and the knowledge thereby produced, then to disciplinary consolidation of knowledge, and finally to something like the Bretherton diagram, a synthesis of not just the entire planet but multiple ways of knowing the planet.

Interdisciplinarity can take many forms. The Bretherton diagram represented the integration of knowledge from across many scientific disciplines. It also gestured toward the need for integrating the social sciences as a means of modeling the role of human factors in the system. The awareness of planetary change that prompted the series of workshops out of which the diagram emerged also prompted a call for another flavor of interdisciplinarity, this time for bringing the science of climate and the traditional discipline of history together.

This awareness was prompted directly by ice cores. Thanks to the magic of isotope chemistry and Dansgaard's "one really good idea," ice cores became frozen annals. The earth, it turned out, not only had a history but had kept its own, remarkably detailed, archive. Ice cores, just one member of a remarkable family of paleo-proxies, stand out for their ability to register very long time spans with remarkably high resolution. Indeed, it is possible in some ice cores to read the earth's history on an annual basis, as a historian might read a church register. This special feature of ice cores made it possible to align human history and climate history in a way never before possible. The earth now had a history that could be directly calibrated to human history. This raised questions, foremost among them how climate affected human history. It was precisely to "evaluate the effects of climate and weather on human affairs in the past" that a 1979 conference on "Climate and History" held at the Climatic Research Unit at UEA brought together 250 researchers from across the sciences, social sciences, and humanities.[2]

At the conference, several things were clear. One was that the question of influence was a complex, multidimensional one. Different human cultures had responded very differently to changes in climate in different times and places. There was no longer any case for the anyway often-derided species of environmental determinism espoused most prominently by Ellsworth Huntington. Also clear was the need for humans to better understand climatic changes in order to prepare for the future. Less clear was whether humans had themselves also influenced climate. In 1979 at UEA, there was no mention of human-induced climate change. The arrow of influence seemed to point decisively from climate to humans, even as scholars emphasized the contingent nature of that influence. There was no room either for environmental determinism or a possible link between human activity and changes in climate. Such a state of affairs did not last long. Increasing evidence about the effects of human activity on rising carbon dioxide in the atmosphere and new indications that global temperatures were rising made the arrow pointing from humans to climate harder to ignore. But in 1979, at the UEA, such concerns could be, and were, pushed to the side.

Ice cores generated a felt need for communication across the aisle that famously separated the two cultures. "'Climate and history,' as a field of study is located at the point of intersection of many different disciplines," asserted the editors of the conference proceedings, "and progress in the field demands interdisciplinary cooperation." History here was a term laden with meanings. "Our approach," explained the editors disingenuously, "is simply to study the history of climate itself, to attempt to reconstruct the pattern of climatic changes and fluctuations over past centuries and millennia."[3] The history of climate, in this sense, could be and often was considered an almost purely scientific endeavor then (save, the editor's note, for a few special historians, Le Roy Ladurie chief among them, who addressed the question of climatic influences on human history).

In asserting that the history of climate was self-evidently scientific, the editors were stating what they believed was an anodyne truth, a mere cliché. But assertions of self-evidence often betray deep uncertainties. Despite the claim to the contrary, there was nothing natural (in the sense of necessary) about the historical nature of climate as represented by these twentieth-century researchers. It was, instead, the product of its own disciplinary history, as contingent and as un-self-evident as any product of humanity.

Geologists like Charles Lyell and James Hutton discovered the earth's "deep time" in the late eighteenth century. But, as Martin Rudwick has convincingly shown, perhaps even more important than the discovery of deep time was the simultaneous creation of a new way of thinking about the earth's past, a new form of historical consciousness that Rudwick calls Earth's "deep history." Much more important for our understanding of the planet than a merely expanded amount of time (what geologist James Hutton famously described as having "no vestige of a beginning and no prospect of an end"), argues Rudwick, was a new sense of "the historicalness or *historicity* of nature."[4] The Bible, with its complex and contingent histories, provided early geologists with a ready model for how change happens over time. From scripture, they borrowed the presiding assumption that the unfolding of the geological history of the earth was a set of contingent events that resembled human history much more than it did the unchanging orbits of the planets described by Isaac Newton.[5] This borrowing from scriptural models of history was itself far from accidental. The sense that at every point events could have turned out differently was, argues Rudwick, "deliberately and knowingly transposed into the world of nature" from that of human culture and human history. Among other things, Rudwick's argument gives lie to the simple conflict stories of science and religion. Far from obstructing the discovery of a deep history of the planet, scriptural understanding "positively facilitated it," argues Rudwick. Though the editors of the UEA conference volume reduced the historicity of climate to something self-evident ("simply . . . the history of climate itself"), geology was "born" historical in a richer and more meaningful sense. Geology was, from its very beginnings, a science built self-consciously on the model of the most human of all histories, that of the Bible.

Climate science today, inasmuch as it has been built partially on the foundations of geology, contains within it some of this historicity, this contingency. But it also contains a different approach to history, one closer in spirit to Newton than to Hutton. The Newtonian time of planetary objects—history that unfolded in precise cycles rather than stories with surprising turns—has always also been a part of what we can now, retrospectively, label climate science. It informed the calculations by which men like James Thomson estimated the melting of ice under pressure. These physical methods gave rise to the kind of thinking that enabled Broecker and others to start working out the global mechanisms responsible for the rapid transformations in the ice records. This work helped create a new way of thinking about the internal history of climate that came to be called climate dynamics.[6] Climate dynamics did not draw explicitly upon traditional historical methods or seek collaboration, as the scientists at the 1979 UEA conference did, with traditional historians in building time lines. Nevertheless, it was a new way of thinking about the climate. Unlike those geologically indebted disciplines which were more or less content to simply describe the unfolding of climate history (for example, classical climatology and many aspects of meteorology and oceanography before the postwar period), scientists interested in physical dynamics wanted to generate a causal understanding of how the pieces of a system connected and how those connections generated phenomena that could be measured. Doing climate history here entailed understanding the causal relations between physical phenomena rather than "merely" describing them. According to this way of thinking, the movements of water, air, or ice had histories that could be generated not through observation and description alone but through the correct application of physical principles. Henry Stommel's paper on the westward intensification of boundary currents is the *locus classicus* for this sort of thinking within oceanography. It not only captured the drive toward understanding physical phenomena that lay at the heart of this kind of climate history, but demonstrated the value of simplicity in this new arena.

In this sense, scientists who studied the dynamics of climate were also self-consciously historical. While there was always uncertainty about the precise path that the climate system followed—a real sense that things could easily have unfolded differently—the science of climate dynamics emphasized not this uncertainty, but the links between elements of the system. In other words, they were more concerned with what could be explained causally, in terms of physical dynamics, and less with what was—at least theoretically—fundamentally unpredictable. In that sense, it seems fair to consider them to be historical in their approach, different as it was from the chronological descriptive framework of the classical climatologists. When the uncertainty became more than noise, new theories to account for it had to be developed. Chief among these was the meteorologist Ed Lorenz's description of the chaotic features of certain systems, atmospheric ones in particular. Chaos, as Lorenz understood it, was a way to introduce unpredictability into a system without descending into randomness, "mere" contingency. Chaotic systems are far from random. Instead, they circle around certain stable states while never setting into a fixed rut. But they are unpredictable, confounding the physicist's ambition to make good on the Newtonian promise of perfect knowledge predicated on a keen enough knowledge of initial conditions. Lorenz showed that in chaotic systems, initial conditions could never have been fine enough to preclude the possibility for uncertain outcomes. In exchange for some knowledge, perfect knowledge was ceded.

The search for simplicity is a recurring theme in much scientific effort, not least when facing the confounding complexities of swirling air and water. If the grail of simplicity had a home, it would be a wooden cabin at the tip of Cape Cod, on the campus of the Woods Hole Oceanographic Institution. There, every summer since 1959, a group of scientists have gathered to hash out the simplest ways to describe the motions of fluids on planetary scales. This seminar is called GFD, for geophysical fluid dynamics, and it is an approach to understanding the motions of planetary fluids that has influenced (and been influenced by) much of the science described in this book.[7] It is significant that this conceptually reductive approach to earth science has been nurtured in an un-insulated wooden shack into which can fit no more than two dozen researchers crammed in a motley assemblage of folding chairs and surrounded on three sides by chalk boards. The size of the cabin is important because it has constrained the size of the community. There are not too many scientists who focus on geophysical fluid dynamics, compared, say, to the number of people involved in computer modeling or the even larger numbers of field scientists studying the multifarious aspects of climate. Most have attended the GFD summer school, which has been running for fifty-nine years. The walls of the cabin can remain uninsulated because GFD is only a summer school. It exists from June to August each year and goes quiescent, save for a minimum of administrative functions, until the next year. The aim (and result) is that GFD sits between a variety of disciplines—the very disciplines described in this book. Oceanographers, meteorologists, atmospheric physicists, and glaciologists are among those who apply to study or lecture here. They leave having absorbed a particular way of seeing the planet, which they apply in the course of their doctoral work, postdocs, and subsequent careers.

The GFD seminar emerged out of a joint seminar series hosted in the fall of 1956 by WHOI and MIT. On the MIT side, the seminar was attended by Norman Phillips and Jule Charney, both of whom had recently moved to MIT from the Institute for Advanced Study in Princeton, where Charney had been running the numerical weather prediction work that John von Neumann had championed. Ed Lorenz, a meteorologist, also attended. On the WHOI side, seminar attendees included Henry Stommel, Joanne Malkus, her then-husband Willem Malkus, and Fritz Fuglister (who did the early observations on eddies in the Gulf Stream). Carl-Gustaf Rossby, then visiting WHOI, also participated. Thus a high concentration of the mathematically inclined oceanographic and meteorological community (who might be called, for convenience's sake, theo-reticians) spent two hours together every two weeks, alternating between Woods Hole and Cambridge, MA, not including the din-ner afterward and the car ride between the venues.

At these seminars, a shared language and shared set of interests in understanding the fluid dynamics of the atmosphere and oceans emerged. So too did the idea for a summer school to train graduate students in this way of thinking. In the fall of 1958, George Veronis, Henry Stommel, and Willem Malkus drafted a proposal for a GFD summer school on the topic of "Theoretical studies in geophysical hydrodynamics." Both Joanne Malkus and Henry Stommel were early advisors, though Joanne stopped attending after her divorce from Willem Malkus, who remained closely involved. Stommel attended for several years. Both exemplified the ethos of GFD. They sought physical insights into the motions of air and water that could explain the complexities of the world in the simplest terms possible.

The first program consisted of four students and six invited staff, in addition to WHOI members. Rather than teaching a set curricu-lum, the program consisted largely in seminars describing the work currently being tackled by the staff. Questions were not merely tol-erated but encouraged, and the emphasis was not so much on con-veying a set body of knowledge but on students and staff together exploring interesting research questions. Stommel and Alan Rob-inson talked about their recently developed theory of the so-called thermocline, or strata of the ocean in which the temperature dropped dramatically. Joanne Malkus gave a talk on cloud physics.

The presiding ethos was of equality. This egalitarian spirit remains as strong as ever as the seminar completes its sixtieth anniversary, with constructive interruptions welcomed at the seminars and a spirit of constant questioning that breaks down barriers between students and faculty.

✳

The impact of the GFD seminar on shaping our understanding of how oceans, ice, and atmospheres move has been large. But the history of climate science has been just as much a story of increasing complexity as it has been of the simplified visions of the GFDers. As important as the Bretherton diagram was, it has long since been surpassed in importance by another kind of global vision in which the values of both simplicity and complexity can be inscribed. It is this global vision—even more than the glamorous image of our "blue marble" against the inky black of space—which has been responsible for shaping how we think about the earth's climate. This is the General Circulation Model (GCM), a complex simulation that tries to reproduce the dynamics of the earth system by calculating how a grid of data points respond to a set of physical equations. Like Jorge Luis Borges's ironically "perfected" art of cartography in which a map of an empire occupies the entirety of an empire, these GCMs aim to cover the globe as completely as possible. Instead of paper, they use imaginary grids whose resolution improves with each rise in computer processing power.[8] Time is another factor in climate models. While it would be possible to run a more highly spatially resolved climate model by taking bigger increments of time, scientists have generally used time steps of thirty minutes to run models over a century or more. This translates into 1,753,152 steps at each of the grid points of any given model. For each grid point, a series of what are called model parameters—values for temperature, wind speed, pressure, humidity, and so on—would also need to be calculated. Multiplying these three sets of numbers by each other—the number of time steps, the number of points on a grid, and the number of values for each point—quickly generates an almost unimaginably huge number of calculations. For the most finely resolved

GCMs currently in use, the number of calculations needed to run the model for a century taxes even the fastest and most powerful computers in the world. As a rule, doubling the resolution of the model results in ten times more calculations being required.[9] Like thirsty behemoths, these models suck up all the available computing power with each advance of Moore's famous law.

GCMs have had notable success in reproducing certain aspects of the climate system, such as the great ocean and atmospheric currents, the pulsing growth and decay of ice caps, and the distribution of atmospheric carbon dioxide. Other features—particularly those operating on small spatial or temporal scales—are harder to capture, even using the largest computers available. The resolution of these kinds of GCMs is currently some 100 kilometers. Anything smaller—a cloud or a small ocean eddy—gets missed out. (Since clouds are key aspects of the global climate system, scientists have worked hard to find other ways of including them. They do this by parameterizing—by finding mathematical shorthands for summarizing the effects of clouds. These are useful tools and better than ignoring such small-scale features altogether, but they are also limited.) The complexity of these model worlds (and there are dozens of them, to complicate matters even more) is such that climate scientists now worry that they are in danger of forgetting that they are not studying the actual earth but a model version of it. Getting lost in the byways of a GCM, they are at risk of losing sight of the real aim of these models—to understand our own planet.[10]

There are other climate models that sit at the other end of an imaginary spectrum of models. These simple models seek not so much to simulate climate as to provide a useful medium for exploring it. A good example of one of these is the energy–balance model of the kind Joanne Simpson and Herbert Riehl used to "discover" hot towers. By eliminating as much detail as possible, these models obey an opposite epistemology to the GCMs. Taking away as much as can be taken away to leave the essential features of the climate system intact offers a powerful kind of vision. This tradition is an old one—stretching back to the work of men such as Croll, Ferrell, and James Thomson. It often has a counterfactual quality, playing around not with the aim of approximating the earth but of imagining alternative earths. Oceanographer John Marshall's aqua planet models do this well.[11] Marshall asks what the earth's climate would be like if its entire surface were covered with water. Letting this model spin out through 5,000 years, he finds that it eventually falls into a settled climate regime—an ice cap forms on both of the poles. Marshall runs the experiment four times, each time adding a line to represent the simplest possible landmass, which serves to interrupt the flow of water around the planet. With four simple variations, Marshall is able to test the importance of landmass distribution on ocean circulation and climate regime—to better understand whether a planet will experience fixed ice ages, oscillating ice ages, or descend into a permanent snowball state.

In theory, between simple models like these aqua planets and complex GCMs lies a series of intermediate models of increasing complexity. The climate system is so complex, according to those who promote the so-called "hierarchy of models" approach, that we need a system of nested models to understand the many scales on which energy flows through the system. The "answer" to the big questions of climate science, according to this view, lies not in any particular model but in the different understandings that each of these hierarchically arranged models enable.[12]

Today, climate scientists are intensely self-conscious about not only their disciplinary but also their epistemic identities—how they know what they know. At conferences which aim for interdisciplinary thinking, it is normal for scientists to preface comments by saying "as a modeler" or "as a theoretician," and so forth. Such thinking was also present, in a different guise, in the dispute between Tyndall and Forbes over the nature of glacier motion. Walker struggled with the limitations of statistics to generate physical understanding. And so too was it present in the concerns of Stommel and Simpson to find the correct balance between observing the complex phenomena of the ocean and atmosphere and finding ways to describe it using the special pithiness of math and physics. The interplay between people "doing" observing, people "doing" theory, and people "doing" modeling has been a central theme to this book. A self-consciousness of the need for balance (where what counts as balance is itself a moving target), rather than a recipe for precise ratios, is a consistent feature of the 150 years of history covered here.

While it is tempting to assert an increasing mathematization of the earth sciences—along the lines of GFD—it seems more accurate to say instead that the need for iteration between theory, observation, and modeling has intensified, and the cycle has sped up. Theorists need data, as much if not more of it than they ever did. And those who generate data—through observations or through modeling—need theory with which to shape their research focus and even, as Paul Edwards has convincingly explained, to see their data at all.

Historians tend to be jumpy about the risks of something that goes by the name of presentism. The tendency to see the past in light of the present is seen as a Bad Thing—blinding us to the truth of the past by seeing it with our foreign eyes. But presentism is inevitable. We cannot escape from the perspective from which we view the past—that is, right now. Rather than struggling to deny this perspective, we need to face it head-on. In light of the environmental challenges facing the world today, we urgently need to think hard about the relationship between the present and the past. Any fears about how we are blinded by our present prejudices seem increasingly less significant than the risk of depriving ourselves of the best tools we can use for imagining the future.

The uses of history to imagine possible futures can go by many names. The past is sometimes looked to as a source of lessons, or case studies, like the climate analogues sought by paleoclimate scientists or old weather patterns used to make new forecasts. We can learn from equivalent moments in the past and (the implicit suggestion seems to be) avoid making the same mistakes. A somewhat more nuanced approach seeks to use the past not as a cheat sheet for future events but as an exercise in imaginative stretching—a way to prepare us to see differently, to use the difference of the past to help us conceive of the future with more options in mind. Anticipatory history is one term for this, developed by those engaged in thinking about how to manage real things—often heritage sites—whose location in the landscape makes them literally susceptible to imminent change.[13] This forces the mind to focus on the issue when it might otherwise skirt it. When it comes to history of climate, the problem and the application are less clear—our scientific practice doesn't feel under threat by climate change in the same way that our landscape does.

But if we think about this harder, perhaps science is under threat. Not, I think, only from those who seek to undermine its authority to speak, though this threat is real and stubbornly hard to dispel. Instead, climate science may be at risk from a lack of self-consciousness. What climate science seems to need is a vocabulary for making explicit what are usually implicit assumptions about the values that inform it. There are many such values embedded in the doing of science, but the historian in me would like to make a special plea for examining the nature of the histories implicit in our view of climate.[14] What, for example, counts as history in climate? Which tools, both conceptual and material, are used to generate such histories? What moments do they render meaningful, and which are ignored or erased? These questions—which we've only just begun to ask—are essential to determining what we care about when it comes to climate, and the answers to them will form the basis—whether we realize it or not—of our responses to the changes we face.

The historical senses that are embedded in climate help determine what counts as normal when it comes to climate. Determining what is a natural climate is a key focus within climate science and policy today. As we learn more about how the climate has changed in the past and consider how it might change in the future, we rely on assumptions about what a "good" or "natural" climate might be. These assumptions have so far been defined by those who study the earth's past. These are not historians but scientists who look to past records of climate change to determine both what we can expect and what we should be happy to accept. How much change is acceptable may be a partly scientific question, but it is not only that. What counts as acceptable changes depends signally on where you draw your frame.[15] The past 12,000 years of history, called the Holocene, have been, in the context of the preceding millions, both unusually stable and unusually warm. This happens to be the period in which human beings evolved. Do we have a responsibility to maintain this particular climate? There are even more versions of normal if we broaden our population of those who have the right to determine the answer to this question beyond climate scientists alone.

Stating that there are many different kinds of knowledge would seem to be a recipe only for dispute of the partisan kind that has caused a crisis of trust, or in some cases an outright rejection of the values of science. Science, in this point of view, is under threat and must be saved by its own methods—by proving with evidence that it "works," which is to say that it can make meaningful predictions. In another sense, understanding the many-strandedness of science can offer a way toward understanding something even more fundamental—the limitations of science. To acknowledge the limits of science need not be an exercise in abnegation. It can open up new ways forward. Identifying the presence and necessity of various values within science, such as the importance of interestedness, commitment, emotional connection, and self-determination, provides a better understanding of what science is. This clears the way to recognizing that the decisions which we make as a society about how we live on the planet can be informed by scientific values without being determined by them. Our choices about how we use energy, how we dispose of our goods, how we live with and in the landscape, have always been about so much more than, for example, our understanding of the ice ages, or our ability to predict the weather.

The nature of the relationship of climate and history has become both a key political issue and an unresolved scientific question. The earth is now self-evidently a planet of change. The past is now always a resource for the future. Scientists now seek to understand the dynamics of the climate in the past, based on paleo records, in order to better understand how it might change in the future. Politically, the question of what futures are available to us—of what futures we imagine—is also partly constrained by this scientific understanding of climate dynamics in the past and future.

The dance goes on. The Bretherton diagram, influential as it once was, now looks dated and clunky. From a focus on understanding the components of the earth "system," as the NASA engineers considered it, a new way of looking at the planet emphasizes not the boxes, so to speak, but the arrows between them.[16] Feedback loops and their associated tipping points have come to the fore. The artificiality of distinctions between elements of the system have given way to a new sense that there can be no substantive distinctions between aspects of the whole. It is all connected in such deep and complex ways that only by studying the connections can any sense of it be made. Some may argue that a study of connections implies also a study of discrete elements, but the emphasis seems to have shifted. The "essential oneness" which Victor Starr drew attention to some sixty years ago is a recurrent theme, but one which, it seems, every generation must arrive at independently.

The scientists whose work I've described in this book were each, in their own ways, playful in their approaches to knowledge—they treated the planet as an arena for exploration. Walker's throwing of the boomerang is the most literal form of play described here, and all the more striking against the austere backdrop of his mathematical intensity. Tyndall was playful, too, in ways that were always testing—his own appetite for risk, the forbearance of his peers with his tendency to dispute, the delicacy of his apparatus as he asked it to answer increasingly difficult questions. Piazzi Smyth played with authority—his own, his instruments', and what we might now call the "truthiness" of images. In seeking knowledge in the most evanescent of phenomena, he played with his own desire for understanding, challenging himself ultimately to accept on faith what he could not prove. Joanne Simpson played in the arena of the sky, using whatever tools she needed—from airplanes to hand-calculated models to her own cloud photographs—to get to the physics she sought. Henry Stommel was playful in his making and his thinking, tinkering with things as he tinkered with ideas, using his mind as a telephoto lens that zoomed in and out, and across the oceanic landscape, seeking problems he thought worth thinking about. For Willi Dansgaard, the icy landscape of Greenland hid a frozen past across which his mind could wander at will, thanks to his one "really good idea."

Work, for these individuals, was a quest, at once playful and completely serious, that took place across decades and landscapes both mental and physical. Following water, and the heat it held (or held traces of), they drew trajectories through time and space just as the molecules they studied did. Their playful exploring was, in its seeking, searching quality, elevated by a poignant sense of longing—for more knowledge, more time with which to study the planet, more freedom in their work, and more tools with which to see deeply. Play, for these individuals, was an avenue for something serious, something big. They each, in their own way, sought something deeply meaningful from their engagement with the planet. So should we all.

ACKNOWLEDGMENTS

I could not have written this book without the expertise and generosity of many people. Speaking with scientists who study water in the oceans, water vapor in the atmosphere, and ice in glaciers and ice sheets has been one of the principal pleasures of writing it. I am happy for the opportunity to thank these generous people here. David Marshall enthusiastically shared his knowledge and loaned me several all-important but hard-to-find volumes from Stommel's *Collected Works*. Carl Wunsch read and commented on several chapters, sharing his historical sensitivity and mastery of the field. During a 2017 visit to Woods Hole Oceanographic Institution, Rui Xin Huang shared his memories of Henry Stommel and made sure I understood something of the special culture of the place, including the GFD seminar. At Woods Hole, Joe Pedlosky and John Marshall also spent time talking with me about both the history and current state of physical oceanography. Back in the UK, Giles Harrison welcomed me to Reading, shared his work on atmospheric physics, and inducted me into the joyful practice of balloon launchings.

I am grateful to the following people for reading and commenting on draft chapters: George Adamson, Matthias Heymann, Mike Hulme, Peggy LeMone, David Marshall, Richard Staley, Spencer Weart, Ed Zipser, and the participants of the following seminars: "Towards a History of Paleoclimatology: Changing Roles and Shifting Scales in Climate Sciences," a workshop held at the Centre for

Environmental Humanities at the University of Hamburg, September 6–7, 2017; "Estimated Truths: Water, Science and Politics of Approximation," held at the Max Planck Institute for the History of Science, August 16–17, 2017; and a summer school on "History of Physics: Scientific Instruments and Environmental Physics" convened by the St. Cross Centre for the History and Philosophy of Physics Centre at Brasenose College, Oxford, August 20–24, 2018. I also benefited from conversations and emails with Karen Aplin, Wallace Broecker, Harry Bryden, Ian Hewitt, Jim Ledwell, Martin Mahony, Dennis Moore, Walter Munk, Chris Rapley, Emily Shuckburgh, John Tennyson, Chris Wilson, and two anonymous reviewers for the University of Chicago Press. All remaining errors are mine alone. In addition, I want to thank Dave Sherman at the Data Library and Archives of the Woods Hole Oceanographic Institution; Diana Carey at the Schlesinger Library, Radcliffe Institute; Karen Moran at the Library of the Royal Observatory Edinburgh; and the staff at the Institute Archives and Special Collections, MIT, the Royal Society in London, and the Bodleian Library in Oxford.

I am grateful to have Peter Tallack of the Science Factory as my agent, and have benefited from two superb editors: Karen Merikangas Darling, at the University of Chicago Press, and Philip Gwyn Jones, at Scribe UK. This book is much the better for their thoughtful and enthusiastic engagement. It is also better thanks to the support of a 2015–2016 Public Scholar grant from the National Endowment for the Humanities.

Friends who have helped keep me afloat include Hayley MacGregor, Sylvie Zannier-Betts, Signe Gosmann, Liz Woolley, Laura Stark, Patrick Tripp, and Susie Reiss. Thanks to all. Now we can finally talk about something else.

My parents, Paul and Cecie Dry, and my sister, Katie Dry, have always supported me, and this book is no exception. I am grateful for their unconditional love and patience with a project that overflowed all of the deadlines I set for it.

At home, I am lucky to have two special people: Jacob, who buoys me like no other, and Rob, who always believes in me.

NOTES

CHAPTER 1

1   Tyndall Centre for Climate Change Research at the University of East Anglia; Geoffrey Cantor, Gowan Dawson, James Elwick, Bernard Lightman, and Michael S. Reidy, eds., *The Correspondence of John Tyndall* (London: Pickering and Chatto, 2014–); and Roland Jackson, *The Ascent of John Tyndall: Victorian Scientist, Mountaineer, and Public Intellectual* (Oxford: Oxford University Press, 2018).

2   Stephen Schneider, "Editorial for the First Issue of *Climatic Change*," *Climatic Change* 1, no. 1 (1977): 3–4.

3   John Tyndall, *The Forms of Water in Clouds and Rivers, Ice and Glaciers* (London: King, 1872), 6.

4   Simon Schama, *Landscape and Memory* (London: HarperCollins, 1995), 7–9.

5   See, for example, Sheila Jasanoff, "Image and Imagination: The Formation of Global Environmental Consciousness," in P. Edwards and C. Miller, eds., *Changing the Atmosphere: Expert Knowledge and Environmental Governance* (Cambridge, MA: MIT Press, 2001), 309–337. For a longer history of global images, see Dennis Cosgrove, *Apollo's Eye: A Cartographic Genealogy of the Earth in the Western Imagination* (Baltimore: Johns Hopkins University Press, 2001); and Sebastian Grevsmühl, *La Terre vue d'en haut: l'invention de l'environnement global* (Paris: Editions du Seuil, 2014).

CHAPTER 2

1   See John Tyndall, "Winter Expedition to the Mer de Glace, 1859," in *The Glaciers of the Alps: being a narrative of excursions and ascents, an account of the origin and phenomena of glaciers and an exposition of the physical principles to which they are related* (London: John Murray, 1860), 195–218. For this expedition, Tyndall employed Eduard Simond and Joseph Tairraz as guides, and four additional porters (199). See also the typescript journals of John Tyndall, vol. 3, section

8, 24–30 December 1859, 101–175, held by the Royal Institution; and Jackson, *Tyndall*, chapter 8, "Storms over Glaciers, 1858–1860," 149–150.

2   Tyndall, *Journals*, vol. 3, 101.

3   Tyndall, *Glaciers*, 208.

4   Tyndall, *Journals*, vol. 3, 159.

5   For more on the dispute between Tyndall and Forbes, see J. S. Rowlinson, "The Theory of Glaciers," *Notes and Records of the Royal Society of London* 26 (1971): 189–204; Bruce Hevly, "The Heroic Science of Glacier Motion," *Osiris* 11 (1996): 66–86: and Jackson, *Tyndall*, chapter 8, "Storms over Glaciers, 1858–1860," 132–151.

6   John Tyndall, "On the Physical Phenomena of Glaciers," *Philosophical Transactions* 149 (1859): 261–278.

7   Martin Rudwick, *Worlds Before Adam: The Reconstruction of Geohistory in the Age of Reform* (Chicago: University of Chicago Press, 2008); and Martin Rudwick, *Earth's Deep History: How It Was Discovered and Why It Matters* (Chicago: University of Chicago Press, 2014).

8   Cited in Christopher Hamlin, "James Geikie, James Croll, and the Eventful Ice Age," *Annals of Science* 39 (1982): 569.

9   Crosbie Smith and Norton Wise, *Energy and Empire: A Biographical Study of Lord Kelvin* (Cambridge: Cambridge University Press, 1989), 556.

10   Rudwick, *Earth's Deep History*, 150.

11   William Hopkins, "On the Causes which may have produced changes in the Earth's superficial temperature," *Quarterly Journal of the Geological Society* 8 (1 February 1852): 88.

12   In 1851, Hopkins read a paper to the Geological Society, in which he quoted Poisson's estimate that only about one-twentieth of a degree of the mean temperature of the earth was due to the so-called "primitive heat." Not only was this percentage small, it was diminishing at such a slow rate that it would take "a hundred thousand millions of years" to reduce this fraction by half. Even for those geologists accustomed to requiring vast amounts of time for the changes of the earth, this was a long time. See Crosbie Smith, "William Hopkins and the Shaping of Dynamical Geology: 1830–1860," *British Journal for the History of Science* 22, no. 1 (March 1989): 41.

13   Hopkins, "On the Causes," 59. Hopkins noted that while previously geologists could only imagine "changes of climatal conditions" from a "higher to a lower general temperature on the earth's surface," more "accurate geological research" had shown that "these changes have been to a considerable extent of an oscillatory character," and "so far as they may be thus characterized, they cannot of course be accounted for by the earth's internal heat."

14   From James Campbell Irons, *Autobiographical Sketch of James Croll, with Memoir of his Life and Work* (London: Edward Stanford, 1896), 32. For more on Croll, see James Fleming, "James Croll in Context: The Encounter between Climate Dynamics and Geology in the Second Half of the Nineteenth Century," *History of Meteorology* 3 (2006): 43–54.

15   Irons, *Croll*, 35.

16   Irons, *Croll*, 228.

17   Cited in Fleming, "Croll," 49.

18   Hamlin, "Geikie," 580.

19   Herschel to Lyell, 6 February 1965; Herschel to Lyell, 15 February 1865, both in Herschel Papers.

20 Charles Darwin to James Croll, 19 September 1868, cited in Irons, *Croll*, 200.

21 James Geikie, *The Great Ice Age and Its Relation to the Antiquity of Man* (London: W. Isbister, 1874), 94.

22 Hamlin, "Geikie," 578.

23 Geikie, *Great Ice Age*, 95.

24 Irons, *Croll*, 104.

25 John Tyndall to James Croll, 14 January 1865, cited in Irons, *Croll*, 104.

26 Tyndall, *Forms*, 7.

27 Tyndall, *Forms*, 14.

28 "Glacial Theories," *North American Review* 96, no. 198 (January 1863): 2.

29 See Crosbie Smith, "William Thomson and the Creation of Thermodynamics: 1840–1855," *Archive for the Exact Sciences* 16 (1977): 231–288.

30 William Hopkins, "On the Theory of the Motion of Glaciers," *Philosophical Transactions of the Royal Society* 152 (1862): 677.

31 See Naomi Oreskes and Ronald Doel, "The Physics and Chemistry of the Earth," in Mary Jo Nye, ed., *The Cambridge History of Science* (Cambridge: University of Cambridge Press, 2003), 544.

32 Hevly, "Heroic Science"; and Michael Reidy, "Mountaineering, Masculinity, and the Male Body in Victorian Britain," in Robert Nye and Erika Milam, eds., "Scientific Masculinities," *Osiris* 30 (November 2015): 158–181.

33 Tyndall, *Glaciers*.

34 Tyndall, *Glaciers*, v.

35 Tyndall, *Glaciers*, 116.

36 Cited in Daniel Brown, *The Poetry of Victorian Scientists: Style, Science and Nonsense* (Cambridge: Cambridge University Press, 2013), 110.

37 Cited in Brown, *Poetry*, 117.

38 Cited in Rowlinson, "Theory," 194.

39 Tyndall, "The Bakerian Lecture: On the Absorption and Radiation of Heat by Gases and Vapours, and on the Physical Connexion of Radiation, Absorption and Conduction," *Philosophical Transactions of the Royal Society* 151 (1861): 1.

40 The device and the challenges it posed are described in Tyndall, "Bakerian Lecture."

41 Diary of John Tyndall, summer 1861, Royal Institution.

42 Diary of John Tyndall, 18 May 1859, Royal Institution.

43 At first, Tyndall hadn't even bothered to test water vapor and carbon dioxide; since they existed in such small quantities in the atmosphere, he assumed that "their effect upon radiant heat must be quite inappreciable." A. J. Meadows, "Tyndall as a Physicist," in W. H. Brock, N. D. McMillan, and R. C. Mollan, eds., *John Tyndall: Essays on a Natural Philosopher* (Dublin: Royal Dublin Society, 2918), 88. Citation from John Tyndall, *Heat Considered as a Mode of Motion* (London: Longmans, Green and Company, 1863), 333.

44 Tyndall, "Bakerian Lecture," 6.

45 Tyndall, "Bakerian Lecture," 29.

46 Tyndall, "Bakerian Lecture," 28.

47 On Tyndall and Magnus, see Jackson, *Tyndall*, 166–168.

48 John Tyndall, "On the Relation of Radiant Heat to Aqueous Vapour," *Philosophical Transactions of the Royal Society of London* 153 (1863): 1–12, at 10.

49 A. S. Eve and C. H. Creasey, *Life and Work of John Tyndall* (London: Macmillan, 1945).

50   Tyndall, *Glaciers*, 205.
51   Tyndall, *Glaciers*, 206.
52   Tyndall, *Glaciers*, 205.

CHAPTER 3

1   Charles Piazzi Smyth, *Teneriffe, An Astronomer's Experiment, Or, Specialties of a Residence Above the Clouds* (London: Lovell Reeve, 1858).
2   Alexander von Humboldt and Aimé Bonpland, *Personal Narrative of Travels to the Equinoctial Regions of the New Continent During the Years 1799–1804* (London: Longman Hurst, 1814), 110.
3   Charles Darwin, *A Naturalist's Voyage: Journal of Researches into the Natural History and Geology of the Countries Visited during the Voyage of H.M.S. Beagle Round the World: Under the Commands of Capt. Fitz Roy, R.N.* (London: John Murray, 1889), 1.
4   Quoted in Kurt Badt, *John Constable's Clouds* (London: Routledge and Kegan Paul, 1950), 55.
5   For biographical information on Piazzi Smyth, see Hermann Brück and Mary Brück, *The Peripatetic Astronomer: The Life of Charles Piazzi Smyth* (Bristol and Philadelphia: Adam Hilger, 1988). For Piazzi Smyth's role in the visual and popular culture of Victorian meteorology, see Katharine Anderson, *Predicting the Weather: Victorians and the Science of Meteorology* (Chicago: University of Chicago Press, 2005), chapter 5; and Katharine Anderson, "Looking at the Sky: The Visual Context of Victorian Meteorology," *British Journal for the History of Science* 36, no. 3 (2003): 301–332.
6   Agnes Clerke, *A Popular History of Astronomy during the Nineteenth Century* (London: Adam & Charles Black, 1893), 152.
7   Simon Schaffer, "Astronomers Mark Time: Discipline and the Personal Equation," *Science in Context* 2, no. 1 (1988): 115–145.
8   Stephen Case, "Land-Marks of the Universe: John Herschel against the Background of Positional Astronomy," *Annals of Science* 72, no. 4 (2015): 417–434.
9   Humboldt and Bonpland, *Personal Narrative*, 110.
10   Humboldt and Bonpland, *Personal Narrative*, 182–183.
11   Alexander von Humboldt, *Cosmos: A Sketch of a Physical Description of the Universe*, trans. E. C. Otte (New York: Harper, 1858), 26.
12   Humboldt, *Cosmos*, 37.
13   Alexander von Humboldt, "Beobachtungen über das Gesetz der Wärmeabnahme in den höhern Regionen der Atmosphäre, und über die untern Gränzen des ewigen Schnees," *Annalen der Physik* 24 (1806): 1–2.
14   Michael Dettelbach, "The Face of Nature: Precise Measurement, Mapping, and Sensibility in the Work of Alexander von Humboldt," *Studies in the History and Philosophy of Science* 30, no. 4 (1999): 473–504.
15   John Cawood, "The Magnetic Crusade: Science and Politics in Early Victorian Britain," *Isis* 70, no. 4 (1979): 492–518.
16   Clerke, *Popular History*, 177.
17   Piazzi Smyth, *Teneriffe*, 77.
18   Piazzi Smyth, *Teneriffe*, 90.

19   Charles Piazzi Smyth, "The Ascent of Teneriffe," *Literary Gazette and Journal of Belles Lettres, Science, and Art*, 17 April 1858 (London: Lovell Reed): 377.

20   Piazzi Smyth, *Teneriffe*, 108–109.

21   Charles Piazzi Smyth, *Astronomical Observations Made at the Royal Observatory Edinburgh* (Edinburgh: Neill and Company, 1863), 444.

22   Piazzi Smyth, *Teneriffe*, 274.

23   Piazzi Smyth, *Teneriffe*, 320.

24   Piazzi Smyth, *Teneriffe*, 288.

25   Charles Piazzi Smyth, "On Astronomical Drawing," *Memoirs of the Royal Astronomical Society* 15 (1946): 75–76.

26   Charles Babbage, *Reflexions on the Decline of Science in England* (London: B. Fellowes, 1830), 210–211.

27   Charles Piazzi Smyth, *Our Inheritance in the Great Pyramid* (London: Alexander Strahan, 1864).

28   Brück and Brück, *Peripatetic Astronomer*, 119.

29   Brück and Brück, *Peripatetic Astronomer*, 177.

30   David Brewster and J. H. Gladstone, "On the Lines of the Solar Spectrum," *Philosophical Transactions of the Royal Society of Edinburgh* 150 (1860): 152.

31   See Anderson, *Predicting*, chapter 5, for a discussion of Piazzi Smyth's rainband spectroscopy. Charles Piazzi Smyth, "Spectroscopic Weather Discussions," *Nature* 26 (5 October 1882): 553.

32   F. W. Cory, "The Spectroscope as an Aid to Forecasting Weather," *Quarterly Journal of the Royal Meteorological Society* 9, no. 48 (1883): 285.

33   Piazzi Smyth, "Spectroscopic," 553.

34   Charles Piazzi Smyth, "The Spectroscope and the Weather," *Popular Science* 22 (1882): 242.

35   Piazzi Smyth, "Spectroscopic," 552.

36   Robert Multhauf, "The Introduction of Self-Registering Meteorological Instruments," *Contributions from the Museum of History and Technology: Paper 23, United States National Museum Bulletin* (Washington, DC: Smithsonian, 1961).

37   Robert H. Scott, *Instructions for the Use of Meteorological Instruments* (London: J. D. Potter, 1875), 9–10.

38   Robert Brain and M. Norton Wise, "Muscles and Engines: Indicator Diagrams and Helmholtz's Graphical Methods," in *Universalgenie Helmholtz: Rückblick nach 100 Jahren*, ed. Lorenz Krüger (Berlin: Akademie-Verlag, 1994), 124–145; and Lorraine Daston and Peter Galison, "The Image of Objectivity," *Representations* 40 (1992): 81–128.

39   Lorraine Daston, "Cloud Physiognomy," *Representations* 136, no. 1 (Summer 2016): 45–71; and Richard Hamblyn, *The Invention of Clouds: How an Amateur Meteorologist Forged the Language of the Skies* (London: Picador, 2001).

40   Ralph Abercromby, *Seas and Skies in Many Latitudes, Or Wanderings in Search of Weather* (London: Edward Stanford, 1888).

41   William Clement Ley, *Cloudland: A Study on the Structure and Character of Clouds* (London: Edward Stanford, 1894), vii.

42   "Manchester Photograpic Society," *British Journal of Photography* (22 December 1876): 609.

43   Brück and Brück, *Peripatetic Astronomer*, 217.

44   Charles Piazzi Smyth, *Cloud Forms That Have Been at Clova, Ripon, 1892–1895*, 3 vols., Archives of the Royal Society.

45 H. H. Hildebrandsson and Teisserenc de Bort, *International Cloud Atlas* (Paris: 1896), 15.

46 Piazzi Smyth, introductory note, *Cloud Forms*, 5, 7.

CHAPTER 4

1 Biographical sources on Walker include S. K. Banerji, "Sir Gilbert Walker CSI, ScD, FRS," *Indian Journal of Meteorology and Geophysics* 10, no. 1 (1959): 113–117; Geoffrey Taylor, "Gilbert Thomas Walker, 1868–1958," *Biographical Memoirs of Fellows of the Royal Society* 8 (November 1962): 166–174; J. M. Walker, "Pen Portrait of Gilbert Walker, CSI, MA ScD, FRS," *Weather* 52, no. 7 (1997): 217–220; and, on Walker's work with the Indian Meteorological Department, D. R. Sikka, "The Role of the India Meteorological Department, 1875–1947," in Uma Das Gupta, ed., *Science and Modern India: An Institutional History, c. 1784–1947*, 381–421, vol. 15, part 4, of D. P. Chattopadhyaya, ed., *History of Science, Philosophy and Culture in Indian Civilization* (Delhi: Pearson-Longman). For a sustained treatment of Walker's work on the Southern Oscillation, see Richard Grove and George Adamson, *El Niño in World History* (London: Palgrave Macmillan, 2018), especially chapter 5, "The Discovery of ENSO," 107–137; and Mike Davis, *Late Victorian Holocausts: El Niño Famines and the Making of the Third World* (London: Verso, 2002), especially part 3, "Decyphering El Niño," 211–239.

2 *The Queen's Empire: A Pictorial and Descriptive Record, Illustrated from Photographs*, vol. 2 (London: Cassell, 1897–1899), 120.

3 Frederik Nebeker, *Calculating the Weather: Meteorology in the 20th Century* (San Diego, CA: Academic Press, 1995), 197n21.

4 Nebeker, *Calculating the Weather*, 21.

5 Deborah Coen, "Climate and Circulation in Imperial Austria," *Journal of Modern History* 82, no. 4 (2010): 846.

6 Julius von Hann, *Handbook of Climatology*, trans. Robert De Courcey Ward (New York: Macmillan, 1903), 2.

7 Climate scientists today still use these techniques, and the assumptions they share with Hann (that it is useful to consider some periods of climate as fundamentally stable) continue to shape the way we understand climate. See Mike Hulme, Suraje Dessai, Irene Lorenzoni, and Donald Nelson, "Unstable Climates: Exploring the Statistical and Social Constructions of 'Normal' Climate," *Geoforum* 40 (2009): 197–206.

8 Hann, *Handbook*, 2.

9 See Coen, "Climate," 846; and Deborah Coen, *Climate in Motion: Science, Empire, and the Problem of Scale* (Chicago: University of Chicago Press, 2018), 139–143.

10 "Notes from India," *Lancet* 157, no. 4045 (15 June 1901): 1713.

11 Mike Davis, *Late Victorian Holocausts: El Niño Famines and the Making of the Third World* (London: Verso, 2002), 26.

12 Davis, *Late Victorian Holocausts*, 32.

13 Davis, *Late Victorian Holocausts*, 146.

14 Report of the Indian Famine Commission 1880, part 1 (Parliamentary Paper, c. 2591), vol. 52 (1881): 25.

15   *Times of India*, 11 June 1902.

16   Davis, *Late Victorian Holocausts*, 152–155.

17   Biographical material on Walker is from Geoffrey Taylor, "Gilbert Thomas Walker, 1868–1958," *Biographical Memoirs of Fellows of the Royal Society* 8 (1962): 166–174; and Walker, "Pen Portrait."

18   Cited in Taylor, "Walker," 168.

19   Letter from Cleveland Abbe to Gilbert Walker, February 24, 1902, in Gilbert Walker Papers, Science Museum Library Archive, MS2012/39.

20   Frank Cundall, *Reminiscences of the Colonial and Indian Exhibition* (London: William Clowes & Sons, 1886), 116.

21   Katharine Anderson, *Predicting the Weather: Victorians and the Science of Meteorology* (Chicago: University of Chicago Press, 2005), 260–261.

22   Norman Lockyer, "Sunspots and Famines," *Nineteenth Century* 2, no. 9 (1877): 583–602.

23   *Imperial Gazetteer of India*, chapter 3, "Meteorology" (London: Clarendon Press, 1909), 104.

24   Norman Lockyer, "The Meteorology of the Future," *Nature* 8 (12 December 1872): 99.

25   Cited in J. Norman Lockyer and W. W. Hunter, "Sunspots and Famine," *Nineteenth Century* (1877): 591.

26   Clerke, *Popular History*, 176. See also Helge Kragh, "The Rise and Fall of Cosmical Physics: Notes for a History, c. 1850–1920," https://arxiv.org/abs/1304.3890, accessed 17 December 2018.

27   Balfour Stewart and Norman Lockyer, "The Sun as a Type of the Material Universe," *Macmillan's Magazine* 18, no. 106 (August 1868): 319–327, at 327.

28   Lockyer and Hunter, "Sunspots and Famine," 585.

29   Lockyer and Hunter, "Sunspots and Famine," 602.

30   "Friday August 19, Subsection of Astronomy and Cosmical Physics, Chairman Sir John Eliot," *Report of the Seventy-Fourth Meeting of the British Association for the Advancement of Science Held at Cambridge in August 1904* (London: John Murray, 1905), 456.

31   Report on the Administration of the Indian Meteorological Department from 1907–1908, 7.

32   Eliot, *Report*, 457.

33   Arnold Schuster, "Address to the Belfast Meeting of the British Association for the Advancement of Science," *Report of the Seventy-Second Meeting of the British Association for the Advancement of Science* (London: John Murray, 1902), 519.

34   Cleveland Abbe, *Proceedings of the American Association for the Advancement of Science* 39 (1890): 77.

35   Nebeker, *Calculating the Weather*, 28.

36   Napier Shaw, *Manual of Meteorology* (Cambridge: Cambridge University Press, 1926–1931), 333.

37   Shaw, *Manual of Meteorology*, 333.

38   H. H. Hildebrandsson, "Quelques recherches sur les centres d'action de l'atmosphère," *Kungliga Svenska Vetenskapsakademiens Handlingar* 29 (1897); Teisserenc de Bort, "Etude sur les causes qui determinant la circulation de l'atmosphère"; and H. F. Blanford, "On the Barometric See-Saw between Russia and India in the Sun-Spot Cycle," *Nature* 25 (1880): 447–482.

39   H. H. Hildebrandsson and Teisserenc de Bort, *Atlas International des nuages: pub conformenent aux decisions du Comite meteorologique international* (Paris: Gauthier-Villars, 1896).

40   Gilbert Walker, "World Weather," *Quarterly Journal of the Royal Meteorological Society* 54 (April 1928): 226.

41   Gilbert Walker, "Correlation in Seasonal Variation of Weather, VIII: A Preliminary Study of World Weather," *Memoirs of the Indian Meteorological Department* 24 (1923): 75–131, 109.

42   Walker, "Correlation," 109.

43   Gilbert Walker, "On Periods and Symmetry Points in Pressure as Aids to Forecasting," *Quarterly Journal of the Royal Meteorological Society* 72, no. 314 (1946): 265–283.

44   Eliot, *Report*, 453.

45   Gilbert Walker, "Seasonal Foreshadowing," *Quarterly Journal of the Royal Meteorological Society* 56 (237): 359–364.

46   Charles Daubeny, *Climate: An Inquiry into the causes of its differences and into its influence on vegetable life, comprising the substance of four lectures delivered before the Natural History society, at the museum, Torquay, in February 1863* (London and Oxford: John Henry and James Parker, 1863).

47   As Normand, the man who headed the Met Office from 1927 to 1944, wrote, "On the whole, Walker's world-wide survey ended by offering more promise for the prediction of events in other regions than in India." Charles Normand, "Monsoon Seasonal Forecasting," *Quarterly Journal of the Royal Meteorological Society* 79 (October 1953): 469.

48   Gilbert Walker, "Presidential Address to the Fifth Indian Science Congress, Lahore, January 1918," *Journal and Proceedings of the Asiatic Society of Bengal, New Series Vol. XIV, 1918* (Calcutta: Asiatic Society): lxxvii.

49   Nebeker, *Calculating the Weather,* 48.

50   From R. B. Montgomery, "Report on the Work of GT Walker," *Monthly Weather Review* 39 (1940): supplement 1–22.

51   Sikka, "The Role," 397.

52   Sikka, "The Role," 401.

53   Sikka, "The Role," 415.

54   Sikka, "The Role," 401.

55   J. Bjerknes, "Atmospheric Teleconnections from the Equatorial Pacific," *Monthly Weather Review* 97 (1969): 163–172.

CHAPTER 5

1   William Koelsch, "From Geo- to Physical Science: Meteorology and the American University, 1919–1945," in *Historical Essays on Meteorology, 1919–1995: The Diamond Anniversary History Volume of the American Meteorological Society,* ed. James Fleming (Boston: American Meteorological Society, 1996), 541–556.

2   Cited in Robert Marc Friedman, "Constituting the Polar Front, 1919–1920," *Isis* 73, no. 3 (September 1982): 355.

3   See Roger Turner, "Teaching the Weather Cadet Generation: Aviation, Pedagogy, and Aspirations to a Universal Meteorology in America, 1920–1950,"

in *Intimate Universality: Local and Global Themes in the History of Weather and Climate*, ed. James R. Fleming, Vládimir Jankovic, and Deborah R. Coen (Sagamore Beach, MA: Science History Publications, 2006), 141–173.

4 Joanne Malkus, "Large-Scale Interactions," in *The Sea: Ideas and Observations on Progress in the Study of the Seas*, vol. 1, *Physical Oceanography*, ed. M. N. Hill (New York: Wiley Interscience, 1962), 99.

5 W.-K. Tao, J. Halverson, M. LeMone, R. Adler, M. Garstang, R. House Jr., R. Pielke Sr., and W. Woodley, "The Research of Dr Joanne Simpson: Fifty Years Investigating Hurricanes, Tropical Clouds, and Cloud Systems," *AMS Meteorological Monographs* 29, no. 15 (January 2003): 1.

6 Duncan Blanchard, "The Life and Science of Alfred H. Woodcock," *BAMS* 65, no. 5 (1984): 460.

7 Bergeron later recognized the limits of his own experience, noting that "I then hardly had seen any weather or climate south of 50 degrees N (except the winter of 1928/29 on Malta)." On Bergeron, see Robert Marc Friedman, *Appropriating the Weather: Vilhelm Bjerknes and the Construction of a Modern Meteorology* (Ithaca, NY: Cornell University Press, 1989); Roscoe Braham, "Formation of Rain: A Historical Perspective," in *Historical Essays on Meteorology, 1919–1995*, 181–223; and Arnt Eliassen, "The Life and Science of Tor Bergeron," *Bulletin of the American Meteorological Society* 59, no. 4 (April 1978): 387–392.

8 Herbert Riehl, "Preface," *Tropical Meteorology* (New York: McGraw-Hill, 1954).

9 Alfred Woodcock and J. Wyman, "Convective Motion in Air over the Sea," *Annals of the New York Academy of Sciences* 48 (1947): 749–776.

10 Michael Garstang and David Fitzjarrald, *Observations of Surface to Atmosphere Interactions in the Tropics* (New York: Oxford University Press, 1999), 58.

11 "Interview with Joanne Simpson," in *The Bulletin Interviews*, ed. Hessam Taba (Geneva: WMO, 1988), 271.

12 See Blanchard, "Woodcock," 460; and "American Meteorological Society, University Corporation for Atmospheric Research, Tape Recorded Interview Project, Interview of Joanne Simpson, 6 September 1989, Interviewer Margaret LeMone" (hereafter Simpson Oral History), 21, in Papers of Joanne Simpson, 1890–2010, Schlesinger Library, Radcliffe Institute (hereafter Simpson Papers).

13 Simpson Papers, MC 779, Simpson 1.13, Family History Overview, Childhood, 2.

14 Simpson Oral History, 21.

15 Simpson Papers, MC 779, Simpson 1.8, Notes between Simpson and lover C, 1950s.

16 J. S. Malkus, "Some Results of a Trade-Cumulus Cloud Investigation," *Journal of Meteorology* 11 (1954): 220–237.

17 Simpson Papers, MC 779, 1.4, Simpson letter re: self-hypnosis for migraines, January 1996.

18 Simpson Papers, MC 779, 1.4, Journal re: Simpson and lover "C" 1952–54, 1 of 2. Entry dated 16 October 1952.

19 Simpson Papers, MC 779, Simpson 2.10, Beginnings of a research career, 1953–1964.

20 J. S. Malkus, "Some Results," 220–237.

21 Simpson Papers, MC 779, 2.10, Summary of the Meteorological Activities of Joanne S. Malkus year 1954–55, Clippings, Beginnings of a research career.

22 See Blanchard, "Woodcock," 460.

23   Henry Stommel, "Entrainment of Air into a Cumulus Cloud," *Journal of Meteorology* 4 (June 1947): 91–94.

24   Deborah Coen, "Big Is a Thing of the Past: Climate Change and Methodology in the History of Ideas," *Journal of the History of Ideas* (April 2016): 305–321.

25   Victor Starr, "The Physical Basis for the General Circulation," in *Compendium of Meteorology*, ed. Thomas Malone (American Meteorological Society, 1951), 541.

26   Robert Serafin, "The Evolution of Atmospheric Measurement Systems," in *Historical Essays on Meteorology, 1919–1995.* During the war, some eighty of these were deployed daily across the United States, and the numbers continued to grow after the hostilities ended.

27   Carl-Gustaf Rossby, "The Scientific Basis of Modern Meteorology," in *Climate and Man, Yearbook of Agriculture* (Washington, DC: U.S. Department of Agriculture, 1941), 599–655.

28   *New York Times,* 11 January 1946, 12.

29   Philip Thompson, "The Maturing of the Science," *Bulletin of the American Meteorological Society* 68, no. 6 (June 1987): 631–637.

30   "If the super-calculator could be built and operated successfully, weather experts said, it not only might lift the veil from previously undisclosed mysteries connected with the science of weather forecasting." *New York Times,* 11 January 1946, 12.

31   "Weather to Order," *New York Times,* 1 February 1947.

32   John von Neumann, "Can We Survive Technology?," in *Fabulous Future: America in 1980* (New York: Dutton, 1956), 152.

33   Von Neumann, "Can We," 108, 152.

34   "Weather to Order."

35   "Making Weather to Order," *New York Times,* 20 July 1947.

36   "Weather to Order."

37   See chapter 12, "The Unification of Meteorology," in Nebeker, *Calculating the Weather.*

38   "Making Weather to Order."

39   Jule Charney, "Impact of Computers on Meteorology," *Computer Physics Communications* 3 (1972 Suppl.): 124.

40   See David Atlas and Margaret LeMone, "Joanne Simpson 1923–2010," *Memorial Tributes: National Academy of Engineering* 15 (2011): 368–375; W.-K. Tao et al., "Research," 4.

41   Nebeker, *Calculating the Weather,* 124; Jacob Bjerknes, "Practical Application of H. Jeffrey's Theory of the General Circulation," *Résumé des Mémoires Réunion d'Oslo* (1948): 13–14; and Victor Starr, "An Essay on the General Circulation of the Earth's Atmosphere," *Journal of Meteorology* 5 (1948): 39–43.

42   Herbert Riehl and Joanne Malkus, "On the Heat Balance in the Equatorial Trough Zone," *Geophysica* 6, no. 3–4 (1958): 534.

43   They cautioned their readers to remember that "many of the quantities are to be based on calculation as residuals rather than independent measurement and are therefore subject to a considerable margin of error." Riehl and Malkus, "On the Heat Balance," 505.

44   Simpson Papers, MC779, Simpson 3.10, Joanne Simpson Notebooks on Research II: Second Set April 1957–July 1959, Evolution of hot towers hypothesis, 1.

45   Malkus, "Large-Scale Intentions," 95.

46   "The gaps in at least our gross factual information are currently being removed rather rapidly." Starr, "Physical Basis," 541.

47   Malkus, "Some Results"; Joanne Starr Malkus and Claude Ronne, "On the Structure of Some Cumulonimbus Clouds Which Penetrated the High Tropical Atmosphere," *Tellus* 6 (1954): 351–366; Joanne Starr Malkus, "On the Structure of the Trade-Wind Moist Layer," *Papers in Physical Oceanography And Meteorology* 12, no. 2 (1958): 47.

48   See, for example, Herbert Riehl, "On the Role of the Tropics in the General Circulation," *Tellus* 2 (1951): 1–17; Herbert Riehl, *Tropical Meteorology* (New York: McGraw-Hill, 1954), chapters 3 and 12; Herbert Riehl, "General Atmospheric Circulation of the Tropics," *Science* 135 (1962): 13–22; and Riehl and Malkus, "On the Heat Balance."

49   Starr, "Physical Basis," 549.

50   For an early review of this work, see Herbert Riehl and Dave Fultz, "Jet Stream and Long Waves in a Steady Rotating-Dishpan Experiment: Structure of the Circulation," *Quarterly Journal of the Royal Meteorological Society* (April 1957): 215–231; and Oral History Interview with Dave Fultz, http://n2t.net/ark:/85065 /d7ks6pzf.

51   H. E. Willoughby, D. P. Jorgensen, R. A. Black, and S. L. Rosenthal, "Project Stormfury: A Scientific Chronicle, 1962–1983," *Bulletin American Meteorological Society* 66, no. 5 (May 1985): 505.

52   Roger Revelle and Hans Suess, "Carbon Dioxide Exchange Between Atmosphere and Ocean and the Question of an Increase of Atmospheric CO_2 during the Past Decades," *Tellus* 9, no. 1 (February 1957): 18–27.

53   Revelle and Suess, "Carbon Dioxide," 20.

54   Richard Anthes, "Hot Towers and Hurricanes: Early Observations, Theories and Models," in Wei-Kuo Tao, ed., *Cloud Systems, Hurricanes and the Tropical Rainfall Measuring Mission (TRMM): A Tribute to Joanne Simpson* (Boston: American Meteorological Society, 2003), 139.

55   Simpson Oral History, 14.

56   Simpson Papers, MC 779, Simpson 3.10, Malkus-Riehl collaboration and Notebooks, 8–10, 10.

57   Joanne Malkus and Herbert Riehl, "On the Dynamics and Energy Transformations in Steady-State Hurricanes," *Tellus* 12, no. 1 (1960): 1–20; and Herbert Riehl and Joanne Malkus, "Some Aspects of Hurricane Daisy, 1958," *Tellus B* 12, no. 2 (May 1961): 181–213.

58   Simpson Papers, MC 779, Simpson 2.8, Scrapbook on clips, 1947–1973, "Head in clouds, mind on weather," *LA Times*, 1961.

59   Simpson Oral History, 11.

60   Simpson Papers, MC 779, Simpson 3.12, Narrative The Miami Years, 1967–1974, 7; and Simpson Oral History, 15.

61   Simpson Papers, MC 779, Simpson 3.12, Stormfury Cumulus Seeding Experiments—Joanne's model tests, Narrative The Miami Years, 1967–1974.

62   Simpson Papers, MC 779, Simpson 3.12, Decade of Weather Modification Experiments, 1964–1974, 8.

63   Simpson Papers, MC 779, Simpson 3.12, Narrative The Miami Years, 1967–1974.

64   Simpson Papers, MC 779, Simpson 3.12, Stormfury Cumulus Seeding Experiments—Joanne's model tests, Narrative The Miami Years, 1967–1974, 9.

65 "We are now able to perform actual experiments in a full-scale atmospheric laboratory in order to evolve and test various modification hypotheses." From Robert Simpson and Joanne Malkus, "Experiments in Hurricane Modification," *Scientific American* 211, no. 6 (1964): 37; and "Seeded Clouds 'Explode,'" *Science News-Letter* 86, no. 8 (1964): 115.

66 Simpson and Malkus, "Experiments," 35.

67 John Walsh, "Weather Modification: NAS Panel Report and New Program Approved by Congress Reveal Split on Policy," *Science* 147, no. 3655 (15 January 1965): 276; and "Weather and Climate Modification: Report of the Special Commission on Weather Modification," National Science Foundation and Advisory Committee on Weather Control, Final Report I, 1957.

68 Cited in Arthur Schlesinger, *A Thousand Days: John F. Kennedy in the White House* (New York: Houghton Mifflin Harcourt, 2002), 910.

69 Simpson Papers, MC 779, Simpson 3.12, Stormfury Cumulus Seeding Experiments—Joanne's model tests, Narrative The Miami Years, 1967–1974, 9.

70 NAS Report on Weather and Climate Modification—Problems and Prospects, NAS-NRC 1350 (Washington, DC: National Academy of Sciences–National Research Council, 1966), 6.

71 NAS Report, 8.

72 NAS Report, 9.

73 NAS Report, 10.

74 Simpson Papers, MC 779, Simpson 3.12, Stormfury Cumulus Seeding Experiments—Joanne's model tests, Narrative The Miami Years, 1967–1974, 14.

75 Simpson Papers, MC 779, Simpson 4.9, Banquet talk, 4 October 1989, Joanne Simpson, AMS President, "The Weather Modification Paradox Rises Again."

76 Simpson Papers, MC 779, Simpson 1.4, Simpson letter re: self-hypnosis for migraines, January 1996.

77 Simpson Papers, MC 779, Simpson 2.10, Clippings, "Woman Cloud Expert Has Time for Family," 2 May 1953, *Boston Evening Globe*. See also "Scientist with Her Feet on Cloud 9," *LA Times*, 20 December 1963.

78 "Woman Likes to Fly in Hurricane's Eye," *Boston Globe*, 1957.

79 "It was then as it is now that no article is published on the work of a female scientists without the details of her spouse, children and home being part of it. That is just fine with me; it is too bad that it is done so rarely when the work of male scientists is discussed in the media." Simpson Papers, MC 779, Simpson 210, Clippings "Beginning of a research career," 2.

80 Simpson Papers, MC 779, Simpson 1.14, Family history narrative, Personal memories January 1996 Re: difficult childhood, depression, referrals to photographs, 1 (unnumbered).

CHAPTER 6

1 George Veronis, "Henry Stommel," *Oceanus* 35 (Special Issue, 1992): 5.

2 Henry Stommel, "The Westward Intensification of Wind-Driven Ocean Currents," *Transactions AGU* 29, no. 2 (April 1948): 202–206.

3 Letter from Iselin to Stommel, 30 April 1950, Woods Hole Oceanographic Institution (WHOI), Papers of Henry Stommel, MC-6, Box 2, Correspondence, 1947–1954.

4   Henry Stommel, *Autobiography*, I-8, in *The Collected Works of Henry Stommel* (Boston: American Meteorological Society, 1995).

5   Stommel, *Collected*, I-9.

6   For biographical information on Stommel, see Arnold Arons, "The Scientific Work of Henry Stommel," in *Evolution of Physical Oceanography: Scientific Surveys in Honor of Henry Stommel*, ed. Bruce A. Warren and Carl Wunsch (Cambridge, MA: MIT Press, 1981); Carl Wunsch, "Henry Melson Stommel: September 27, 1920–January 17, 1992," *National Academy of Sciences Biographical Memoir* 72 (1997): 331–350; and "A Tribute to Henry Stommel," *Oceanus* 35 (Special Issue, 1992). See also Henry Stommel's *Autobiography* in *Collected Works*.

7   Henry Stommel, "Why We Are Oceanographers," *Oceanography* 2, no. 2 (1989): 48–54.

8   Henry Charnock, "Henry Stommel," *Oceanus* 35 (Special Issue, 1992): 15–16.

9   Oliver Ashford, *Prophet or Professor: The Life and Work of Lewis Fry Richardson* (Bristol: Adam Hilger, 1985), 82–83.

10   Henry Stommel, "Response to the Award of the Ewing Medal, from AGU 1977," *Collected*, I-205.

11   L. F. Richardson, "The Supply of Energy from and to Atmospheric Eddies," *Proceedings of the Royal Society* A97 (1920): 354–73.

12   L. F. Richardson and Henry Stommel, "Note on Eddy Diffusion in the Sea," *Journal of Meteorology* 5 (1948): 238–240.

13   Margaret Deacon, *Scientists and the Sea, 1650–1900: A Study of Marine Science* (Aldershot: Ashgate, 1997), 209.

14   Eric Mills, *The Fluid Envelope of Our Planet: How the Study of Ocean Currents Became a Science* (Toronto: University of Toronto Press, 2009), chapter 2; and Deacon, *Scientists,* chapters 14 and 15.

15   Mills, *Fluid Envelope*, 155–158.

16   K. F. Bowden, "The Direct Measurement of Sub-Surface Currents," *Deep Sea Research* 2 (1954): 3–47.

17   B. Helland-Hansen and F. Nansen, *The Norwegian Sea. Its Physical Oceanography Based upon the Norwegian Researches 1900–1904*, Report on Norwegian Fishery and Marine Investigations, vol. 2, part 1 (Bergen: Fiskeridirektoratets, 1909).

18   L. F. Richardson, *Weather Prediction by Numerical Process* (Cambridge: Cambridge University Press, 1922), 66.

19   See, for example: "Once the problem is viewed as a global structural problem in which the Gulf Stream is a limb of a strongly asymmetric circulation cell the problem changes its characters permanently and profoundly. The Gulf Stream is now part of the general circulation of the ocean and not a geographic curiosity." Joe Pedlosky, introduction to chapter 1 of Stommel, *Collected Works*, II-7.

20   Philip Richardson, "WHOI and the Gulf Stream," 2004, at https://www.whoi .edu/75th/book/whoi-richardson.pdf. For more biographical detail, see Jennifer Stone Gaines and Anne D. Halpin, "The Art, Music and Oceanography of Fritz Fuglister," http://woodsholemuseum.org/oldpages/sprtsl/v25zn1-Fuglister.pdf; and "In Memoriam, Valentine Worthington," http://www.whoi.edu/mr/obit /viewArticle.do?id=851&pid=851.

21   F. C. Fuglister and L. V. Worthington, "Some Results of a Multiple Ship Survey of the Gulf Stream," *Tellus* 3 (1951): 1–14.

22   Henry Stommel, "Direct Measurement of Sub-Surface Currents," *Deep Sea Research* 2, no. 4 (1953): 284–285.

23 For biographical information on John Swallow, see Henry Charnock, "John Crossley Swallow, 11 October 1923–3 December 1994," *Biographical Memoirs of Fellows of the Royal Society* 43 (November 1997): 514–519.

24 Henry Stommel, "A Survey of Current Ocean Theory," *Deep Sea Research* 4 (1957): 149–184.

25 John Swallow, "Variable Currents in Mid-Ocean," *Oceanus* 19 (Spring 1976): 18–25.

26 J. C. Swallow and B. V. Hamon, "Some Measurements of Deep Currents in the Eastern North Atlantic," *Deep-Sea Research* 6 (1960): 155–168.

27 J. C. Swallow, "Deep Currents in the Open Ocean," *Oceanus* 7, no. 3 (1961): 2–8; and J. Crease, "Velocity Measurements in the Deep Water of the Western North Atlantic," *Journal of Geophysical Research* 67 (1962): 3173–3176.

28 Stommel notes in his autobiography that by 1950 it was well known that the dynamics of the atmosphere was not linear, and that the possibility of similar dynamics was "always in our minds," but dynamical eddies had not been observed in the ocean until the *Aries* expedition. See Stommel, *Autobiography*, I-39; and Carl Wunsch, "Towards the World Ocean Circulation Experiment and a Bit of Aftermath," in *Physical Oceanography: Developments Since 1950*, ed. Markus Jochum and Raghu Murthugudde (Berlin: Springer, 2006), 182.

29 Henry Stommel, "Varieties of Oceanographic Experience," *Science* 139, no. 3555 (15 February 1963): 575.

30 From Memo of 11 August 1969, by Henry Stommel, Correspondence 1958, 1969–1970, in Mid-Ocean Dynamics Experiment, AC 42 Box 2, Folder 92, MIT Archives.

31 From Memo of 11 August 1969, by Stommel.

32 Stommel, *Collected Works*, I-64.

33 Henry Stommel, "Future Prospects for Physical Oceanography," *Science* 168 (26 June 1970): 1535.

34 Stommel, "Varieties," 572.

35 For more on the Stommel diagram, see Tiffany Vance and Ronald Doel, "Graphical Methods and Cold War Scientific Practice: The Stommel Diagram's Intriguing Journey from the Physical to the Biological Environmental Sciences," *Historical Studies in the Natural Sciences* 40, no. 1 (2010): 1–47. Stommel, "Varieties," 575.

36 They would eventually be joined in the effort by colleagues at Woods Hole, MIT, Harvard, Yale, AOML/NOAA, URI, JHU, Columbia, and Scripps.

37 Stommel, "Future Prospects," 1536.

38 Jochum and Murthugudde, *Physical Oceanography: Developments since 1950*, 51.

39 B. J. Thompson, J. Crease, and John Gould, "The Origins, Development and Conduct of WOCE," in *Ocean Circulation and Climate: Observing and Modelling the Global Ocean,* ed. Gerold Siedler, John Church, and John Gould (San Diego, CA: Academic Press, 2001), 32.

40 *The Turbulent Ocean*, Centre Films, 1974.

41 Allen Hammond, "Physical Oceanography: Big Science, New Technology," *Science* 185, no. 4147 (19 July 1976): 246–247.

42 Hammond, "Physical Oceanography."

43 Stommel, "Why We Are Oceanographers," 50.

44 Henry Stommel, "Theoretical Physical Oceanography," *Collected*, I-119.

45 Francis Bretherton, "Reminiscences of MODE," in *Physical Oceanography: Developments since 1950*, 26.

46   Swallow, "Variable Currents," 24.

47   Peter Rhines, "Physics of Ocean Eddies," *Oceanus* 19, no. 3 (1976): 31.

48   Rhines, "Physics," 35.

49   *The Role of the Ocean in Predicting Climate: A Report of Workshops Conducted by the Study Panel on Ocean Atmosphere Interaction, Under the Auspices of the Ocean Science Committee of the Ocean Affairs Board, Commission on Natural Resources, National Research Council, December 1974* (National Academy of Sciences: Washington, DC, 1974), vi.

50   Stommel, *Collected*, I-217.

51   Stommel, *Collected*, I-72.

52   For a history of the development of CO_2 measurements, see Maria Bohn, "Concentrating on CO_2: The Scandinavian and Arctic Measurements," *Klima Osiris* 26, no. 1 (2011): 165–179.

53   *The Role of the Ocean*, 1.

54   *The Role of the Ocean*, vi.

55   Wunsch, "Towards," 183.

56   Erik Conway, "Drowning in Data: Satellite Oceanography and Information Overload in the Earth Sciences," *Historical Studies in the Physical and Biological Sciences* 37, no. 1 (2006): 134.

57   Wunsch, "Towards," 186–187.

58   "It was obvious that numerical models of the ocean were about to outstrip any observational capability for testing them." In Wunsch, "Towards," 187.

59   Geoff Holland and David Pugh, *Troubled Waters: Ocean Science and Governance* (Cambridge: Cambridge University Press, 2010), 107–108.

60   "The time was ripe to turn again to large-scale oceanography from the process studies which have dominated the attention of oceanographers in recent decades." From foreword, John Mason and R. W. Stewart, vi, World Climate Research Programme, WOCE Scientific Steering Group, Scientific Plan for the World Ocean Circulation Experiment, WCRP Publications Series No. 6, WMO/TD-No. 122, July 1986.

61   J. D. Woods, "The World Ocean Circulation Experiment," *Nature* 314, no. 11 (April 1985): 509.

62   Henry Stommel, "Numerical Models of Ocean Circulation," in proceedings of a symposium held at Durham, NH, 17–20 October 1972, National Academy of Sciences, Washington, DC, 1975, in Stommel, *Collected*, I-202.

63   Woods, "The World," 501.

64   Walter Munk and Carl Wunsch, "Observing the Ocean in the 1990s," *Philosophical Transactions of the Royal Society A* 307 (1982): 440.

65   Stommel, "Why We Are Oceanographers," 52.

66   Stommel, "Why We Are Oceanographers," 54.

67   Interview with Henry Stommel and Bill von Arx, 11 May 1989, Woods Hole Oceanographic Institution Archives.

CHAPTER 7

1   Many of the following biographical details come from Willi Dansgaard's memoir *Frozen Annals: Greenland Ice Cap Research* (Odder, Denmark: Narayana Press, 2004).

2   Willi Dansgaard, "The Abundance of ^{18}O in Atmospheric Water and Water Vapour," *Tellus* 5 (1953): 461-469.

3   Willi Dansgaard, "The ^{18}O Abundance in Fresh Water," *Geochimica et Cosmochimica* 6 (1954): 259.

4   Dansgaard, *Frozen Annals*, 16.

5   Jamie Woodward, *The Ice Age: A Very Short Introduction* (Oxford: Oxford University Press, 2014), 85. See successive editions of James Geikie's *The Great Ice Age* in 1874 and 1877 for evidence of one prominent theorist's shift from a marine to a land ice theory of glaciation.

6   W. B. Wright, *The Quaternary Ice Age* (London: Macmillan, 1937), 74.

7   Cited in James Fleming, *Historical Perspectives on Climate Change* (Oxford: Oxford University Press, 1998), 53. For more on the changes in climate thinking, see Mattias Heymann, "The Evolution of Climate Ideas and Knowledge," *WIREs Climate Change* 1, no. 1 (2010): 588.

8   Cited in John Imbrie and Katherine Palmer Imbrie, *Ice Ages: Solving the Mystery* (Cambridge, MA: Harvard University Press, 1979), 117.

9   Dansgaard, "^{18}O Abundance."

10   On the history of the WMO and for a brief review of previous attempts at international meteorology, see Paul Edwards, "Meteorology as Infrastructural Globalism," *Osiris* 21 (2006): 229-250.

11   On the history of the IAEA-WMO collaboration, see P. K. Aggarwal et al., "Global Hydrological Isotope Data and Data Networks," in J. West, G. Bowen, T. Dawson, and K. Tu, eds., *Isoscapes* (Dordrecht: Springer, 2010), 33-50.

12   Willi Dansgaard, "Stable Isotopes in Precipitation," *Tellus* 16 (1964): 437.

13   Roger Launius, James Fleming, and David DeVorkin, *Globalizing Polar Science: Reconsidering the International Polar and Geophysical Years* (Basingstoke: Palgrave, 2011); and Ronald Doel, Robert Marc Friedman, Julia Lajus, Sverker Sörlin, and Urban Wråkberg, "Strategic Arctic Science: National Interests in Building Natural Knowledge through the Cold War," *Journal of Historical Geography* 44 (2014): 60-80.

14   Cited in Janet Martin-Nielsen, "'The Deepest and Most Rewarding Hole Ever Drilled: Ice Cores in the Cold War in Greenland," *Annals of Science* 70 (2012): 56.

15   On Camp Century and ice-core drilling, see Edmund Wright, "CRREL's First 25 Years, 1961-1986" (CRREL, 1986), 1-65; Chester Langway Jr., *The History of Early Polar Ice Cores* (U.S. Army Corps of Engineers, 2008); Martin-Nielsen, "'The Deepest'"; and Kristian Nielsen, Henry Nielsen, and Janet Martin-Nielsen, "City under the Ice: The Closed World of Camp Century in Cold War Culture," *Science as Culture* 23 (2014): 443-464. The most extensive treatment of Dansgaard's ice-core research is Maiken Llock, *Klima, kold krig og iskener* (Aarhus: Aarhus University Press, 2006). For a contemporary description of Camp Century, see Walter Wager, *Camp Century: City Under the Ice* (Chilton Books, 1962).

16   See James Fleming, *The Callendar Effect: The Life and Work of Guy Stewart Callendar (1898-1964)* (Boston: American Meteorological Society, Springer, 2007); and Ed Hawkins and Phil Jones, "On Increasing Global Temperatures: 75 Years after Callendar," *Quarterly Journal of the Royal Meteorological Society* 139, no. 677 (2013): 1961-1963.

17   For a full account of the events described in this paragraph, see Spencer Weart, "The Discovery of Global Warming," https://history.aip.org/climate/; for a

much-condensed version of this annually updated online resource, see *The Discovery of Global Warming*, 2nd ed. (Cambridge, MA: Harvard University Press, 2008).

18   Heymann, "The Evolution."

19   Roger Revelle, "Atmospheric Carbon Dioxide," in *Restoring the Quality of Our Environment: Report of the Environmental Pollution Panel, President's Science Advisory Committee* (White House, 1965), 127.

20   Paul Edwards, "History of Climate Modeling," *WIREs Climate Change* 2 (2011): 128–139.

21   Sam Randalls, "History of the 2 Degree Climate Target," *WIREs Climate Change* 1 (2010): 598–605.

22   Paul Edwards, *A Vast Machine: Computer Models, Climate Data, and the Politics of Global Warming* (Cambridge, MA: MIT Press, 2010), 287–322.

23   Mike Hulme, "Problems with Making and Governing Global Kinds of Knowledge," *Global Environmental Change* 20, no. 4 (2010): 558–564.

24   Janet Martin-Nielsen, "Ways of Knowing Climate: Hubert H. Lamb and Climate Research in the UK," *WIREs Climate Change* 6, no. 5 (2015): 465–477.

25   On the broader cultural history of forecasting in America, see Jamie Pietruska, *Looking Forward: Prediction and Uncertainty in Modern America* (Chicago: University of Chicago Press, 2017).

26   On what the ice core revealed, see Willi Dansgaard, S. J. Johnsen, and C. C. Langway Jr., "One Thousand Centuries of Climatic Record from Camp Century on the Greenland Ice Sheet," *Science* 166, no. 3903 (1969): 377–380; and Richard Alley, *The Two-Mile Time Machine: Ice Cores, Abrupt Climate Change, and Our Future* (Princeton, NJ: Princeton University Press, 2000).

27   Dansgaard, Johnsen, and Langway, "One Thousand Centuries," 377–380.

28   On the history of paleoclimatology, see chapter 8 in Woodward, *The Ice Ages*; H. Le Treut, R. Somerville, U. Cubasch, Y. Ding, C. Mauritzen, A. Mokssit, T. Peterson, and M. Prather, "Historical Overview of Climate Change," in *Climate Change 2007: The Physical Science Basis, Contribution of Working Group I to the Fourth Assessment Report of the Intergovernmental Panel on Climate Change*, ed. S. Solomon, D. Qin, M. Manning, Z. Chen, M. Marquis, K. B. Averyt, M. Tignor, and H. L. Miller (Cambridge and New York: Cambridge University Press, 2007); Chris Caseldine, "Conceptions of Time in (Paleo)Climate Science and Some Implications," *WIREs Climate Change* 3 (2012): 329–338; R. W. Fairbridge, "History of Paleoclimatology," in *Encyclopedia of Paleoclimatology and Ancient Environments*, ed. V. Gornitz (New York: Springer, 2009), 414–428; and Matthias Dörries, "Politics, Geological Past, and the Future of Earth," *Historical Social Research* 40, no. 2 (2015): 22–36.

29   Dansgaard, Johnsen, and Langway, "One Thousand Centuries," 380.

30   Spencer Weart, "The Rise of Interdisciplinary Climate Science," *PNAS* 110 (2013): 3658.

31   Wallace Broecker, "Absolute Dating and the Astronomical Theory of Glaciation," *Science* 151 (1966): 299–304.

32   Wallace Broecker, "The Carbon Cycle and Climate Change: Memoirs of My 60 Years in Science," *Geochemical Perspectives* 1 (2012): 276–277; Wallace Broecker, "When Climate Change Predictions Are Right for the Wrong Reasons," *Climatic Change* 142 (2017): 1–6; and Wallace Broecker, *The Great Ocean Conveyor:*

Discovering the Trigger for Abrupt Climate Change (Princeton, NJ: Princeton University Press, 2010), 19–25.

33   From George Kukla, R. K. Matthews, and J. M. Mitchell, "The End of the Present Interglacial," *Quaternary Research* 2, no. 3 (1972): 261–269.

34   For the role played by Soviet climate scientists in the debate over the use of analogues, see Jonathan Oldfield, "Imagining Climates Past, Present and Future: Soviet Contributions to the Science of Anthropogenic Climate Change, 1953–1991," *Journal of Historical Geography* 60 (2018): 41–51.

35   Barry Saltzman, *Dynamical Paleoclimatology: Generalized Theory of Global Climate Change* (San Diego, CA: Academic Press, 2002).

36   Alley, *Two-Mile Time Machine*, 21; and J. Jouzel, "A Brief History of Ice Core Science Over the Last 50 Years," *Climate of the Past Discussions* 9 (3 July 2013): 3711–3767.

37   Author interview, 10 April 2015.

38   More recently, some have suggested other mechanisms to account for the D-O events, such as sea ice feedbacks or tropical processes, and Carl Wunsch has raised the possibility that they represent local or regional changes caused by windfield shifts owing to interaction with the ice sheet, rather than global change. Amy Clement and Larry Peterson, "Mechanisms of Abrupt Global Change of the Last Glacial Period," *Reviews of Geophysics* 46 (2008): 1–39; and Carl Wunsch, "Abrupt Climate Change: An Alternative View," *Quaternary Research* 65 (2006): 191–203.

39   *Global Change: Impacts on Habitability: A Scientific Basis for Assessment: A Report by the Executive Committee of a Workshop held at Woods Hole, Massachusetts, June 21–26, 1982*, submitted on behalf of the executive committee on 7 July 1982 by Richard Goody (Chairman), NASA and Jet Propulsion Lab. See also *Earth Observations from Space: History, Promise, and Reality* (Washington, DC: National Academies Press, 1995).

40   *Global Change*, 3–4.

41   *Toward an Understanding of Global Change: Initial Priorities for US Contributions to the International Geosphere-Biosphere Program* (Washington, DC: National Academies Press, 1988), v.

42   *Earth System Science: A Closer View*, Report of the Earth System Sciences Committee, NASA Advisory Council (Washington, DC: NASA, 1988), 12.

43   The diagram has come to be known as the Bretherton diagram but was developed by Berrien Moore, a future chair of IGBP, according to Sybil Seitzinger et al., "International Geosphere-Biosphere Programme and Earth System Science: Three Decades of Co-Evolution," *Anthropocene* 12 (December 2015): 3–16. Quote from Moore in "Berrien Moore, Earth System Science at 20," Oral History Project, Edited Oral History Transcript, Berrien Moore III, interviewed by Rebecca Wright, National Weather Center, Norman, OK, 4 April 2011.

44   *Earth System Science*, 19.

45   Gregory Good, "The Assembly of Geophysics: Scientific Disciplines as Frameworks of Consensus," *Studies in the History and Philosophy of Modern Physics* 31, no. 3 (2000): 259–292.

46   Sybil P. Seitzinger, Owen Gaffney, Guy Brasseur, Wendy Broadgate, Phillipe Ciais, Martin Claussen, Jan Willem Erisman, Thorsten Kiefer, Christiane Lancelot, Paul S. Monks, Karen Smyth, James Syvitski, and Mitsuo Uematsu, "Inter-

national Geosphere–Biosphere Programme and Earth System Science: Three Decades of Co-Evolution," *Anthropocene* 12 (2015): 3–16.

47   *Earth System Science*, 1.

48   *Earth System Science*, 5.

49   *Earth System Science*, 15 and 10.

CHAPTER 8

1   Juergen Wiechselgartner and Roger Kasperson, "Barriers in the Science-Policy-Practice Interface: Toward a Knowledge-Action-System in Global Environmental Change Research," *Global Environmental Change* 20 (May 2010): 276.

2   H. H. Lamb and M. J. Ingram, "Climate and History: Report on the International Conference on 'Climate and History,' Climatic Research Unit, University of East Anglia, Norwich, England, 8–14 July 1979," *Past & Present* 88, no. 1 (1 August 1980): 137.

3   T. M. L. Wigley, M. J. Ingram, and G. Farmer, eds., *Climate and History: Studies in Past Climates and Their Impact on Man* (Cambridge: Cambridge University Press, 1985), 4.

4   Rudwick, *Earth's Deep History*, 4.

5   Rudwick, *Earth's Deep History*, 4.

6   The origins of this term and group of specialists can be dated by the creation of a journal named *Climate Dynamics* in 1986.

7   For the history of GFD, see George Veronis's very useful informal history of the GFD program at http://www.whoi.edu/page.do?pid=110017.

8   In 1990, when the first IPCC report appeared, the spatial resolution (grid size) was around 500 square kilometers. The grid extends up into the atmosphere as well as horizontally across the earth. Because the atmosphere is so thin compared to the surface area of the planet, it is sliced even more thinly—usually into increments of one kilometer. By 1996, that number had halved, to 250 kilometers. By 2001, it was down to 180 kilometers, and in 2007 it stood at 110.

9   https://eo.ucar.edu/staff/rrussell/climate/modeling/climate_model_resolution.html.

10   Nadir Jeevanjee, "A Perspective on Climate Model Hierarchies," *Journal of Advances in Modeling Earth Systems* 9, no. 4 (August 2017): 1760.

11   See, for example, David Ferreira, John Marshall, Paul O'Gorman, and Sara Seager, "Climate at High-Obliquity," *Icarus* 243 (2014): 236–248.

12   Nadir Jeevanjee, Pedram Hassanzadeh, Spencer Hill, and Aditi Sheshadri, "A Perspective on Climate Model Hierarchies," *Journal of Advances in Modeling Earth Systems* 9, no. 4 (2017): 1760–1771.

13   Caitlin De Silvey, Simon Naylor, and Colin Sackett, eds., *Anticipatory History* (Axminster, Devon: Uniform Books, 2011).

14   One example of this approach is Alessandro Antonello and Mark Carey, "Ice Cores and the Temporalities of the Global Environment," *Environmental Humanities* 9, no. 2 (2017): 181–203.

15   See Mike Hulme, Suraje Dessai, Irene Lorenzoni, and Donald Nelson, "Unstable Climates: Exploring the Statistical and Social Constructions of 'Normal' Climate," *Geoforum* 40 (2009): 197–206.

16   See, for example, Gisli Palsson, Bronislaw Szerszynski, Sverker Sörlin, John
Marks, Bernard Avril, Carole Crumley, Heide Hackmann, Poul Holm, John
Ingram, Alan Kirman, Mercedes Pardo Buendía, and Rifka Weehuizen, "Recon-
ceptualizing the 'Anthropos' in the Anthropocene: Integrating the Social
Sciences and Humanities in Global Environmental Change Research," *Environ-
mental Science & Policy* 28 (2013): 3–13.

BIBLIOGRAPHIC ESSAY

CHAPTER 1

Classic studies of global images of Earth are Tim Ingold's "Globes and Spheres: The Topology of Environmentalism," in K. Milton, ed., *Environmentalism: The View from Anthropology* (London: Routledge, 1993), 31–42; and Dennis Cosgrove's "Contested Global Visions: One-World, Whole-Earth, and the *Apollo* Space Photographs," *Annals of the Association of American Geographers* 84 (1994): 270–294. More recent treatments of the nature of global knowledge include Mike Hulme, "Problems with Making and Governing Global Kinds of Knowledge," *Global Environmental Change* 20, no. 4 (2010): 558–564; the special issue on "Visualizing the Global Environmental: New Research Directions," *Geo* 3, no. 2 (2016); Ursula Heise, *Sense of Place and Sense of Planet: The Environmental Imagination of the Global* (Oxford: Oxford University Press, 2008); and Sebastian Grevsmühl, *La Terre vue d'en haut: l'invention de l'environnement global* (Paris: Editions de Seuil, 2014).

CHAPTER 2

Of Tyndall's many works, *Glaciers of the Alps* (London: Murray, 1860), *Heat Considered as a Mode of Motion* (London: Longmans, 1863), and *The Forms of Water in Clouds and Rivers, Ice and Glaciers* (London: King, 1872) are most germane to the topics covered in this chapter. Thanks to Roland

Jackson's recent biography *The Ascent of John Tyndall* (Oxford: Oxford University Press, 2018) and the Tyndall Correspondence Project, which has produced four of a planned nineteen-volume series (published by the University of Pittsburgh Press), it is now possible to dive into Tyndall's private world more easily than ever before. To place Tyndall in his social and cultural context, see Gowan Dawson and Bernard Lightman, eds., *Victorian Scientific Naturalism* (Chicago: University of Chicago Press, 2014); and Bernard Lightman and Michael Reidy, eds., *The Age of Scientific Naturalism: Tyndall and His Contemporaries* (London: Routledge, 2014). The classic article on the relationship between mountaineering, heroism, and science is Bruce Hevly's "The Heroic Science of Glacier Motion," *Osiris* 11 (1996): 66–86. For a more recent discussion of the role of masculinity and mountaineering, see Michael Reidy, "Mountaineering, Masculinity, and the Male Body in Mid-Victorian Britain," in Robert Nye and Erika Milam, eds., "Scientific Masculinities," *Osiris* 30 (November 2015): 158–181.

James Croll still awaits his biographer. He tells his own story in James Campbell Irons, *Autobiographical Sketch of James Croll, with Memoir of His Life and Work* (London: Edward Stanford, 1896). On the development of geology in the period, see Mott Greene, *Geology in the Nineteenth Century: Changing Views of a Changing World* (Cornell, NY: Cornell University Press, 1982), as well as Martin Rudwick's synthesis of his own extensive scholarship on the topic, *Earth's Deep History: How It Was Discovered and Why It Matters* (Chicago: University of Chicago Press, 2014). An excellent history of the idea of the ice ages is John Imbrie and Katherine Palmer Imbrie's *Ice Ages: Solving the Mystery* (Cambridge, MA: Harvard University Press, 1979), as is Jamie Woodward's *The Ice Age: A Very Short Introduction* (Oxford: Oxford University Press, 2014).

CHAPTER 3

Charles Piazzi Smyth's exuberant account of his attempt to prove the feasibility of mountaintop astronomy is titled *Teneriffe, An Astronomer's Experiment: Or, Specialities of a Residence above the Clouds* (London: Lovell Reeve, 1858). To understand his pyramidological obsession, see also *Our Inheritance in the Great Pyramid* (London: Alexander Strahan,

1864), written before he and Jessie visited Egypt, and the three-volume *Life and Work at the Great Pyramid* (Edinburgh: Edmonston and Douglas, 1867) composed upon their return. H. A. Brück and M. T. Brück's biography *The Peripatetic Astronomer: The Life of Charles Piazzi Smyth* (Bristol and Philadelphia: Adam Hilger, 1988) gives a good account of his life but frustratingly lacks footnotes. Katharine Anderson trenchantly analyzes Piazzi Smyth's rainband spectroscopy and cloud photography as part of the visual culture of Victorian meteorology in her *Predicting the Weather: Victorians and the Science of Meteorology* (Chicago: University of Chicago Press, 2005). Larry Schaff places Piazzi Smyth's work at the Great Pyramid and Tenerife in the context of the technical and aesthetic development of photography in a series of articles in *History of Photography*: "Charles Piazzi Smyth's 1865 Conquest of the Great Pyramid," vol. 3, no. 4 (1979): 331–354; "Piazzi Smyth at Tenerife: Part I, the Expedition and the Resulting Book," vol. 4, no. 4 (1980): 289–307; "Piazzi Smyth at Tenerife: Part II, Photography and the Disciplines of Constable and Harding," vol. 5, no. 1 (1981): 27–50. On Piazzi Smyth's role in setting metrical standards, see Simon Schaffer, "Metrology, Metrication and Victorian Values," in *Victorian Science in Context* (Chicago: University of Chicago Press, 1997), 438–474. *Mapping the Spectrum: Techniques of Visual Representation in Research and Teaching* (Oxford: Oxford University Press, 2002), by Klaus Hentschel, explores the astonishing range of epistemological and practical challenges in representing the spectral array. The classic article on the Magnetic Crusade is John Cawood, "The Magnetic Crusade: Science and Politics in Early Victorian Britain," *Isis* 70, no. 4 (1979): 492–518. Successive editions of the International Cloud Atlas from 1896 onward show the evolution of techniques for identifying and ordering cloud types.

On Humboldt, see Andrea Wulf's recent biography *The Invention of Nature: The Adventures of Alexander von Humboldt, Lost Hero of Science* (New York: Knopf, 2015) and dip into the primary texts: Alexander von Humboldt, *Personal Narrative of Travels to the Equinoctial Regions of the New Continent During the Years 1799–1804 by A. von Humboldt and A. Bonpland* (London: Longman Hurst, 1814); and Alexander von Humboldt, *Cosmos: A Sketch of a Physical Description of the Universe*, trans. E. C. Otte (New York: Harper, 1858).

CHAPTER 4

Richard Grove's pioneering *Green Imperialism: Colonial Expansion, Tropical Island Edens and the Origins of Environmentalism* (Cambridge: Cambridge University Press, 1995) provides important context for the relationship between imperial projects and changing understandings of the relations between humans and the environment. To understand the pedagogical system that shaped a young Gilbert Walker, see Andrew Warwick's *Masters of Theory: Cambridge and the Rise of Mathematical Physics* (Oxford: Oxford University Press, 2003), which describes the intellectual and physical rigors of the life of a nineteenth-century Cambridge wrangler. Mike Davis's *Late Victorian Holocausts: El Niño Famines and the Making of the Third World* (London: Verso, 2002) tracks the imperial causes of successive famines in India. In *El Niño and World History* (London: Palgrave Macmillan, 2018), Richard Grove and George Adamson consider El Niño from prehistory to the present.

On the history of sunspots and solar physics, see Graeme Gooday, "Sunspots, Weather and the Unseen Universe: Balfour Stewart's Anti-Materialist Representation of Energy," in *Science Serialized: Representation of the Sciences in Nineteenth Century Periodicals*, ed. Sally Shuttleworth and Geoffrey Cantor (Cambridge, MA: MIT Press, 2004). Deborah Coen's *Climate in Motion: Science, Empire and the Problem of Scale* (Chicago: University of Chicago Press, 2018) outlines the role of the Habsburg monarchy in ushering in a multiscalar science of climate.

CHAPTER 5

Robert Marc Friedman's *Appropriating the Weather: Wilhelm Bjerknes and the Construction of a Modern Meteorology* (Ithaca, NY: Cornell University Press, 1989) tells the history of the Bergen school of meteorology, which combined empirical forecasting with dynamical physics to forge a new meteorology during and in the years immediately following World War I. Frederik Nebeker's *Calculating the Weather: Meteorology in the 20th Century* (San Diego, CA: Academic Press, 1995) describes the growth of meteorology from a broader perspective and across an entire century, including the impact of the computer and the rise of numerical meteorology in the post–World War II period. Kristine Harper's *Weather by the Numbers: The Genesis of Modern Meteorology* (Cambridge, MA: MIT Press, 2008) describes a similar historical period with special insight into the contributions of operational meteorologists to the development of numerical weather prediction. Paul Edwards's *A Vast Machine: Computer Models, Climate Data and the Politics of Global Warming* (Cambridge, MA: MIT Press, 2010) is a masterful analysis of the relations between data, models, and politics in the generation of climate science and provides an important frame for Joanne Simpson's use of models and data in her own work.

On weather control, see Kristine Harper, *Make It Rain: State Control of the Atmosphere in Twentieth-Century America* (Chicago: University of Chicago Press, 2017). Jacob Darwin Hamblin, *Arming Mother Nature: The Birth of Catastrophic Environmentalism* (Oxford: Oxford University Press, 2013), discusses the military uses of weather control, while James Fleming's *Fixing the Sky: The Checkered History of Weather and Climate Control* (New York: Columbia University Press, 2012) is a warning to contemporary would-be geo-engineers of the perils of weather control.

CHAPTER 6

For an overview of the early history of oceanography, see Margaret Deacon, *Scientists and the Sea, 1650–1900: A Study of Marine Science* (Aldershot: Ashgate, 1997). Helen Rozwadowski picks the story up in the mid-nineteenth century and places it in broader cultural context in her *Fathoming the Ocean: The Discovery and Exploration of the Deep Sea* (Cambridge, MA: Harvard University Press, 2005). Eric Mills's *The Fluid Envelope of Our Planet: How the Study of Ocean Currents Became a Science* (Toronto: University of Toronto Press, 2009) presents the history of oceanography as a shift from a descriptive to a physical science, ending at pretty much the precise moment that Stommel entered the scene. Stommel's lively autobiographical memoir is included in the hard-to-find Henry Stommel, Nelson Hogg, and Rui Xin Huang, *Collected Works of Henry M. Stommel*, 3 vols. (Boston: American Meteorological Society, 1995), which contains all of his published and much of his unpublished work. More accessible are the numerous remembrances included in

"Henry Stommel," *Oceanus* 35 (Special Issue, 1992). For the impact of the Stommel diagram outside oceanography, see Tiffany Vance and Ronald Doel, "Graphical Methods and Cold War Scientific Practice: The Stommel Diagram's Intriguing Journey from the Physical to the Biological Environmental Sciences," *Historical Studies in the Natural Sciences* 40, no. 1 (2010): 1–47. *Ocean Circulation and Climate: Observing and Modelling the Global Ocean,* ed. Gerold Siedler, John Church, and John Gould (San Diego, CA: Academic Press, 2001), contains a comprehensive snapshot of the field in 2001, including a retrospective appraisal of MODE. Carl Wunsch's *Modern Observational Physical Oceanography: Understanding the Global Ocean* (Princeton, NJ: Princeton University Press, 2015) demonstrates how observation lies at the heart of physical oceanography today and includes a history of how it came to be.

CHAPTER 7

Willi Dansgaard's lively and funny memoir *Frozen Annals: Greenland Ice Cap Research* (Odder, Denmark: Narayana Press, 2004) paints a vivid picture of the intellectual and practical adventures of the early expeditions to drill ice cores in Greenland. See Ronald Doel, Kristine Harper, and Matthias Heymann's edited volume *Exploring Greenland: Cold War Science and Technology on Ice* (New York: Palgrave Macmillan, 2016) and Janet Martin-Nielsen's *Eismitte in the Scientific Imagination: Knowledge and Politics at the Center of Greenland* (New York: Palgrave Macmillan, 2013) for a good grounding in the strategic importance of Greenland during the Cold War, the importance of geophysics to military defense work in the period, and the part played by a small country, Denmark, in a story normally dominated by the United States and the Soviet Union. Richard Alley's *The Two-Mile Time Machine: Ice Cores, Abrupt Climate Change, and Our Future* (Princeton, NJ: Princeton University Press, 2000) provides an account of ice-core research in Greenland beginning in the 1990s, picking up roughly where Dansgaard's account ends. Wallace Broecker's *The Great Ocean Conveyor: Discovering the Trigger for Abrupt Climate Change* (Princeton, NJ: Princeton University Press, 2010) provides a firsthand account of the role played by paleoclimatological data, including ice cores, in understanding global change. The definitive account of the discovery of global warming remains Spencer Weart's eponymous text (Cambridge, MA: Harvard University Press, 2008, 2nd ed.). For a useful summary of the transition from climatology to climate science, see Matthias Heymann and Dania Achermann, "From Climatology to Climate Science in the Twentieth Century," in S. White, C. Pfister, and F. Mauelshagen, eds., *The Palgrave Handbook of Climate History* (London: Palgrave Macmillan, 2018), 605–632. Joshua Howe's *Behind the Curve: Science and the Politics of Global Warming* (Seattle: University of Washington Press, 2014) is a politically nuanced history. *Earth System Science: A Closer View* (Washington, DC: NASA, 1988) is a fascinating window onto the birth of the discipline.

CHAPTER 8

For a convincing demonstration of how geologists established a historical dimension for nature, see Martin Rudwick's trilogy of books: *Bursting the Limits of Time: The Reconstruction of Geohistory in the Age of Revolution* (Chicago: University of Chicago Press, 2005); *Worlds Before Adam: The Reconstruction of Geohistory in the Age of Reform* (Chicago: University of Chicago Press, 2008); and *Earth's Deep History: How It Was Discovered and Why It Matters* (Chicago: University of Chicago Press, 2014). On the relationship between history and climate, there is a small but growing set of theoretical interventions. See, for example, Dipesh Chakrabarty, "The Climate of History: Four Theses," *Critical Inquiry* 35, no. 2 (2009): 197–222; Fabien Locher and Jean-Baptiste Fressoz, "Modernity's Frail Climate: A Climate History of Environmental Reflexivity," *Critical Inquiry* 38, no. 3 (2012): 579–598; and Andreas Malm, "Who Lit This Fire? Approaching the History of the Fossil Economy," *Critical Historical Studies* 3, no. 2 (2016): 215–248. The concept of the Anthropocene has catalyzed debate on the historical nature of anthropogenic climate change since Paul Crutzen and Eugene Stoermer coined the term in "The 'Anthropocene,'" *IGBP Newsletter* 41 (2000): 17–18. A more recent critical history of the concept is Christophe Bonneuil and Jean-Baptiste Fressoz's *The Shock of the Anthropocene: The Earth, History and Us,* trans. David Fernbach (London: Verso, 2016).

INDEX

Abercromby, Ralph, 105–6, 110
absorption of heat, 1, 48–53, 95, 250
actinometers, 68
Adelaide, 134
Africa, 67–68, 134, 167, 202, 246
Agassiz, Louis, 28–31, 40, 42, 235, 261
Agra, 143
agriculture, 2, 123, 173
airplanes: Dansgaard and, 233, 248; Gerould and, 156–63, 167, 170, 175, 287; hurricanes and, 175, 177; military needs of, 193; ocean studies and, 215; seeding and, 13, 173–86
Airy, George, 68, 71
Alaska, 246
Alpine Club, 26f, 45f
Alps, 17, 23–24, 26, 29, 39–49, 53, 55–56, 272
Antarctica, 15, 27, 221, 258, 263, 271
Anthropocene period, 310n16
anthropology, 2
Apollo 17 mission, 9
Arabian Sea, 133
Arctic, 28, 221, 234–35, 239, 248, 250
Aries (ship), 211
Armstrong, Neil, 182
Arrhenius, Svante, 2, 48, 250, 252
astronomy: Airy and, 68, 71; atmospheric impediments to, 68; Babbage and, 83–84; as cartography, 72; celestial mechanics and, 72, 240; cosmical physics and, 142; experiments and, 82, 86, 93, 95; Gerould and, 148–49; heat and, 80, 91, 94; Herschels and, 33–36, 76, 83–84, 103, 130, 141, 240–41; Humboldt and, 12, 76, 81, 84, 103–4; Kew Observatory and, 102; Leverrier and, 98; Lockyer on, 129; mapping and, 67, 71–72, 117; Nasmyth and, 87–88; Newton and, 72–73, 78, 103, 241; observation and, 68, 70–72, 81, 89, 91; orders

of magnitude and, 81; personal equation and, 70–72; physics and, 97; Piazzi Smyth and, 11, 63–72, 77–104; positional, 72; practical use of, 117; predictive science and, 93, 100, 240–41; Royal Observatory and, 71, 102; as science of patterns, 10; spectrum and, 82–85, 92, 94–104, 107, 112; Stommel and, 193; telescopes and, 79–81, 84, 88–89; Tenerife studies and, 11, 63, 66–70, 73, 77–79, 82, 84–85, 92, 95–96, 107, 130, 142, 272
astrophysics, 274
Atlantic Ocean, 139, 153–54, 173, 176, 190, 194, 201, 204, 206, 210, 212, 217, 263
atlases, 11, 104, 110–11, 114, 139, 142, 202–3, 207, 214
atmosphere: astronomical impediments of, 68; barometers and, 68, 73, 93, 98, 100–102, 106, 110, 129, 133, 136–37, 139 (*see also* pressure); Bretherton diagram and, 265–69, 271, 274, 281, 287; buoyancy and, 168; burning coal and, 3; carbon dioxide and, 3, 5, 51, 60, 168, 174–75, 224–26, 250–53, 258, 275, 282; circulation and, 13, 15, 138, 145, 151, 162, 167–72, 184–85, 213, 216, 222, 224, 226, 241, 247, 251, 278, 281–83; clouds and, 13, 73–85, 89, 92, 95–107, 147–54, 159–86, 293n43, 302n65; General Circulation Model (GCM) and, 281–83; geophysical fluid dynamics and, 279–81; glaciers and, 35, 47, 51–55, 60; Hann and, 118–20; heat and, 3–4, 35, 47, 51–54, 60, 145, 165, 168–69, 227, 237, 250, 252, 293n43; ice and, 232–34, 237, 241–44, 247–53, 258–59, 265; IPCC report and, 309n8; maps of, 167; mathematics and, 115–16; molecules and, 54, 95, 104, 232, 247, 250; monsoons and, 115, 118–19, 126, 128, 133–38, 141–45; NOAA and, 216; nonlinear dynamics of, 304n28; oceans and, 2, 4–6, 13, 15, 35, 78, 126, 133, 142, 145, 162, 168–

319